CREATIVE DINING SPACE

创意餐饮空间设计

（美）艾利克斯·休斯　编
凤凰空间　译

江苏科学技术出版社

图书在版编目（CIP）数据

创意餐饮空间设计 /（美）休斯编 ；凤凰空间译
. -- 南京 ：江苏科学技术出版社，2014.4
ISBN 978-7-5537-2946-6

Ⅰ. ①创… Ⅱ. ①休… ②凤… Ⅲ. ①餐厅－室内装
饰设计－图集 Ⅳ. ①TU247.3-64

中国版本图书馆 CIP 数据核字 (2014) 第 045370 号

创意餐饮空间设计

编　　者	（美）艾利克斯·休斯
译　　者	凤凰空间
项目策划	凤凰空间
责任编辑	刘屹立
出版发行	凤凰出版传媒股份有限公司 江苏科学技术出版社
出版社地址	南京市湖南路1号A楼，邮编：210009
出版社网址	http://www.pspress.cn
总 经 销	天津凤凰空间文化传媒有限公司
总经销网址	http://www.ifengspace.cn
经　　销	全国新华书店
印　　刷	北京建宏印刷有限公司
开　　本	1 016 mm×1 194 mm　1/16
印　　张	23
字　　数	294 000
版　　次	2014年4月第1版
印　　次	2014年4月第1次印刷
标准书号	ISBN 978-7-5537-2946-6
定　　价	328.00元

图书如有印装质量问题，可随时向销售部调换（电话：022-87893668）。

Good restaurant design will be very helpful to the development of restaurant industry and the cultural exchange all over the world. Restaurant industry is a combination of different local features at the different requests of market in the international cities, and it makes the restaurant industry full of diverse colors and of form. Food all over the world adds a lot of color to the local culture. Learning and communicating with each other is an important prerequisite to the development of restaurant industry; meanwhile, it can drive the acculturation. Therefore, the design of restaurant should create comfortable situation by using presenting skills.

This publication is a kaleidoscope of 70 projects, all designed by the most excellent foreign designers. It displays the materials and methods of restaurant design comprehensively. The collection of classical projects will certainly not only bring the readers luxurious amenities of life, but also create a good platform for interactive communication between counterparts.

成功的餐厅空间设计对餐饮业的发展和世界各地的文化交流有莫大的帮助。在当今国际化的城市里，应不同的市场需求，餐饮业融汇了各地不同的特色，使饮食文化多姿多彩。世界各地的美食为地方本身的文化增添了不少色彩，相互学习和交流也是餐饮业发展的重要条件，且能带动整体文化的交融。因此，餐厅空间设计应该运用表现技巧来创造情境和营造氛围。

本书精选了近70个餐厅设计作品，均由优秀的设计师完成，全面展示了餐厅空间设计的方法和材料。经典案例的集结出版定能为读者带来奢华的生活享受，也为同行之间的相互交流提供了良好的平台。

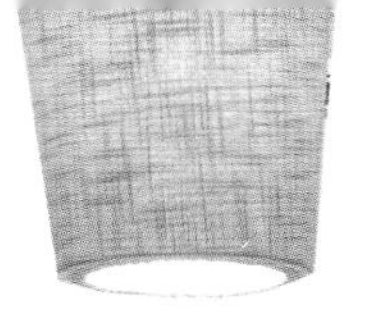

PREFACE

RESTAURANT DESIGN

Restaurant design is a unique craft.
In the field of architecture, restaurant design regularly sits at the very adventurous end of the scale and delivers some of the most flamboyant, unreserved and striking examples of modern design. Restaurant design has developed into an expressive pursuit offering the opportunity to surprise, delight and enthral.
At its core, the very first principal of all restaurant design must be to enhance the commercial advantage of a venue. A restaurantmust survive and thrive as a commercial entity and the elements for success are many, diverse, changeable and nearly impossible to define.
The starting point will be the obvious available space and the opportunities and the constraints of planning, space, services, time, and budget, etc. All are individual criteria for particular circumstances and will be addressed according to practical needs. Moving beyond such design basics as planning, services, space allocation, we shift into the world of the ephemeral – the mood, the atmosphere and the visual message.The successful design of a new venue will complement and showcase the owner's vision and produce, the physical realisation of the owner's dream.
A restaurant is a commercial space in a uniquely capricious marketplace. An owner invests complete reliance on the architect or designer's ability to provide that great space for the presentation and visual branding of his business. The risks are profound. An owner with a clear direction and vision for their new venue entrusts the designer to meet those expectations. The designer's task will be to exceed those expectations to a level never imagined. The design of a restaurant sets the tone of a customer's expectations. A successful design will serve to attract and entice the customer to enter, spend time, enjoy an experience, relate that experienceto others and to return again.
Appropriateness needs to be the greatest design imperative. An appropriate design is vital for the commercial success of a new hospitality business. Appropriateness can be many different things and completely contrary to first expectations. An interior that matches an existing building or indeed the stark opposite, an interior in complete contrast to it's setting. Being polite, being noticed, being audacious, being restrained, being well mannered – all can be appropriate according to the context, the project specifics or the required task.
My own process is to understand the owner's vision and try to visualise how this can be adapted to the available building envelope. Work with the qualities or attributes of a space. Every place will have distinguishing features which can be reflected, enhanced or turned back on themselves to give the first definition, the primary characterisation of a place.
I have been fortunate to work on a wide variety of hospitality projects; from establishing a new restaurant in a 1930's Art Deco dance hall all the way through to converting a former vast crocodile farm into a new 1,200 seat pub/brewery. The most telling and vital lesson learnt has been to be versatile and adapt according to the situation. But always be appropriate to the project requirements at hand.
My personal agenda for restaurant design:
Understand the owner's vision.
Develop a clear concept to present this vision in the design.
Keep everything extremely simple and maintain a stringent clarity of message.
Establish a distinct feature, a design element or a foundation which is likely to become the visual cue for the venue, an enduring image; this can be a component of an existing building or something totally new.
Be consistent with this vision, be bold, be careful.
Continually walk yourself through the design, see the message at the entry, the intrigue beyond, the reasons for staying and the reasons for venturing further.
Provide interest, fascination, charm, outlook and remain mindful of personal comfort.
Know when to stop.
Above all else, aim to produce something enduring. A restaurant with great visual appeal but not so fashionable as to be imminently unfashionable, this is the most difficult and evasive quality of all.
Acknowledge and understand current trends but avoid being captive to them. Look beyond the style of the moment and strive to create something unique and timeless.
Good design can be polite and understated or gregarious and dramatic. But good design will always be consistent with the place it represents and provide an owner with the best possible setting to showcase their talents.
Restaurant design is the imperfect pursuit of the undefinable. It can be intriguing, baffling and demanding. It can also be immensely exhilarating and very gratifying. But the very best of restaurant design illustrates the very best of expressive contemporary architecture.

Paul Burnham Architect

餐厅设计

餐厅设计是一门独特的艺术。
在建筑领域，餐厅设计通常都是很冒险的。曾经也出现过一些浮华、豪放以及引人注目的作品。如今，餐厅设计已经发展成为一件追求惊喜、愉悦以及令人沉醉的事情。
首先，对于所有餐厅设计来说，最重要的原则就是要提高场地的商业优势。一间餐厅必须作为一个商业实体来生存和发展，成功因素不仅多样且多变，更是难以定义。
设计之初要考虑的问题就是明确的可用空间、规划带来的机遇和限制、服务、时间以及预算等。所有这一切都是特定环境下的个别标准，都将根据实际需求而得到解决。超越诸如规划、服务、空间分配等基础设计，我们转移到瞬间世界——情绪、气氛以及视觉信息的设计。新场所的成功设计将完美地诠释和展现雇主的憧憬，并将其梦想转变为现实。
餐厅是多样化市场中的一个商业场所。餐厅所有者完全信任设计师的能力，相信这些设计师能够为餐厅的业务展示和视觉品牌推广提供合理的空间。一个对于自己的餐厅有着清晰方向和明确定位的雇主，相信设计师能够将这些期望转变为现实，而设计师的工作是基于这些期望，创造出超乎想象的设计。餐厅设计同时也要符合客人的期望。一个成功的餐厅设计将有助于吸引客人步入其中，令其愿意在此驻足享受，并且流连忘返。
得体性是最重要的设计规则。一个得体的设计对于一间新餐厅的商业成功至关重要。为了达到设计的得体性，我们需要做很多事情，有些事情与雇主最初的设计期望完全相反。室内装饰可能会与现有的建筑物相匹配或是形成鲜明的对比，或者与周围的环境形成反差。优雅细致、引人注目、大胆创新、严谨克制、儒雅端庄——所有上述的设计风格对于具体的项目以及任务来说都是得体的。
我自己的设计过程是先了解雇主的设计愿望，然后利用专业知识，尝试构思如何将其设计成具有实际可操作性的建筑作品，从而能够与空间的特质和属性相匹配。每一处细节都彰显着回归自我的本色。
我很荣幸地参与了多个餐厅的设计工程，其中包括20世纪30年代极具装饰艺术风格的百乐门餐厅，以及把曾经的鳄鱼养殖场改造成一个新的可以容纳1 200人的餐厅。从中我学到的最有说服力也是最重要的东西就是要根据具体情况因地制宜，同时也要兼顾项目设计的具体要求。
我个人的餐厅设计日程：
理解雇主的设计愿望。
制定一个明确的计划来实现雇主的设计思路。
简化一切程序，并保持清晰的信息脉络。
将一系列独特的设计元素作为设计场所的视觉提示。这个不朽的形象，可以成为现有建筑物或全新场所的一个组成部分。
与最初的设计思路保持一致，小心谨慎，大胆实践。
不断审视自己的设计，查看项目的相关信息，思索原设计规划予以保留或者需要进一步改动的原因。
关注客人的舒适度。
知道在什么时候停止。
除此之外，立志创造一个经久不衰的设计。一间吸引公众眼球的餐厅，但却不是昙花一现，这是最困难、最难以做到的一点。
理解并领悟当前的流行趋势，但同时也要避免程式化。寻找能够超越这种趋势的风格，努力创造独特而永恒的设计。
一个好的设计应该是高雅、含蓄、生动、容易被大众所接受的。与此同时，好的设计也要体现场所的风格，要体现雇主想要传达的理念。
餐厅设计的实质是对于那些难以言明的事物的不完美追求。所以，它可以是有趣的、令人费解的，或者苛刻的；同时，它也可以是令人振奋且愉悦的。然而，最出色的餐厅设计则可以将当代建筑风格展现得淋漓尽致。

设计师：Paul Burnham

On Restaurant Design

Restaurants today have become so much more than a place to eat. They have become a lifestyle choice. Entertainment, a theatre of sorts where you go to be nourished by the food and wine offer, by the service, and the surrounding atmosphere is created by the design of the space. These are the elements, the key components that comprise a restaurant.

Food, eating out and going for a drink or even catching up for a coffee give us the excuse for interaction. Eating is one of the fundamental activities that offer satisfaction on all manner of levels. Eating and drinking spaces can range from the monastically simple to transporting us somewhere beyond the ordinary.

The notion of luxury has been redefined as we have more than ever before. Hence, the definition of luxury has become more about culture, authenticity of product, materials and ingredients. Whether they excite or extinguish your ardour, they are all postcards or snapshots of a time, place and brief but most importantly of this time.

In designing a restaurant, many aspects have to be addressed. How does the concept of the design relate to the concept of the food and beverage offer? Is this easily recognisable and relayed through the design? Is it a comfortable space? Do you want the customers to stay a long time or move quickly on? Is it scaleable? Is it a brand? What is required of the business? Should the restaurant entice, excite and sparkle?

The design of the interior is an integral component of the total equation of a restaurant. Customers want an experience, from not only the food but also the total product and what I call offer.

Being an architect by training and practice and a restaurateur by chance, I have an understanding and appreciation of the infinite and minute detail that goes into the planning and the running and the operational machinations of a restaurant. An interior can be designed to choreograph our consumption whether shopping or eating and drinking. The character of the restaurant should represent the relationship between the space, its food and surrounding community.

关于餐厅设计

如今，餐厅已经不止是一个供客人享受美食的地方。它已经成为一种生活方式的选择——一个享受美食美酒之地，一个感受优质服务与宜人环境之处。以上这些因素共同构成了当今餐厅设计的重要组成部分。

外出就餐、去酒吧喝酒或者去咖啡馆喝咖啡，都给予我们与外界交流的机会。其中，外出享受美食是可以满足所有层次客人的基本活动之一。餐厅的设计风格，有的简朴、大方，有的则高档且别具一格。

本书重新定义了“奢华”的概念，因为相比以往，我们了解了更为全面的信息。因此，对于“奢华”的定义更多的是关于文化、产品、材料和配料的天然真实性。无论是否激发或者熄灭你的热情，它们都是一段时期的缩影，虽然简短，但却至关重要。

在设计一间餐厅时，设计师需要考虑很多因素。如何将餐厅的设计理念与其所要提供的美食、佳酿联系起来？如何设计才会让客人容易辨认，并为后续设计打下基础？如何将这里打造成一个舒适之处？如何留住客人而不是让客人匆匆离去？如何创建一个品牌？作为一桩生意，什么是必须的？一间餐厅应该充满诱惑、激发灵感还是夺人眼目、闪耀发光？

餐厅的内部设计是总体设计构思中一个不可或缺的组成部分，不仅包括食物还包括餐厅所能提供的所有服务。

作为一个身经百战的设计师，同时也是一间餐厅的老板，我对于餐厅设计的细节有着无限且细致的理解，并将这一切贯穿到餐厅的设计和运营之中。餐厅的内部设计无论是在购物还是在就餐方面，都应该符合消费情调。餐厅的特色应该反映出空间、食物以及周围环境的有机结合。

CONTENTS

SHANG PALACE, SHANGRI – LA TAIPEI

Designers: Ed Ng, Terence Ngan
Design Company: AB Concept
Area: 700 m²

A light, regal tone has been set in this series of fine dining spaces. AB Concept chose the popular "celestial cloud" graphic motif to anchor a landscape of contrasting textures and dreamy watercolor fabric shades.
The motif appears first and most memorably behind the maitre d' as a deep sculptural alcove, with its layers lit strikingly from inside, however, it reappears in carpeting, ceiling reliefs and upholstery throughout the restaurant.
The materials are selected for a sense of traditional elegance, for example, brass accents, pearlescent paint finishes and lanterns in delicate silk.
In the private dining spaces, "precious" themes-of pearl, jade and the celestial inform the design schemes; the latter is reserved for the state room, which opens out to its own patio. In the transitional Peony Vestibule, the studio has created another striking impression with a large skylight, where the wooden petal-shaped panels of a ceiling installation soften and diffuse the light.

富丽堂皇是该系列餐厅的特点。AB Concept选择时下流行的“祥云”图案来定位地面的主题风格，即朦胧的水彩底色衬托出交错的纹理。

首先映入眼帘的是位于maitre后面的深色雕刻壁龛，层次由内而外清晰、明朗，令人印象深刻。地毯、天花板和室内装饰材料上，这样的图案也是随处可见。

材料的选择旨在营造一种传统而优雅的氛围，选材中不乏铜黄色的花纹、珠光色的勾边和精美丝绸上的灯饰。

包房区的“珠宝”主题就餐区——珍珠厅、翡翠厅、祥云厅彰显着本店的设计特色。后者是高级包房，面对着酒店的中庭。在传统的牡丹前庭中，AB Concept设计了开阔的天窗，给人留下深刻的印象。这里层层叠叠的花瓣状实木天花板使原本强烈的日光温和了许多。

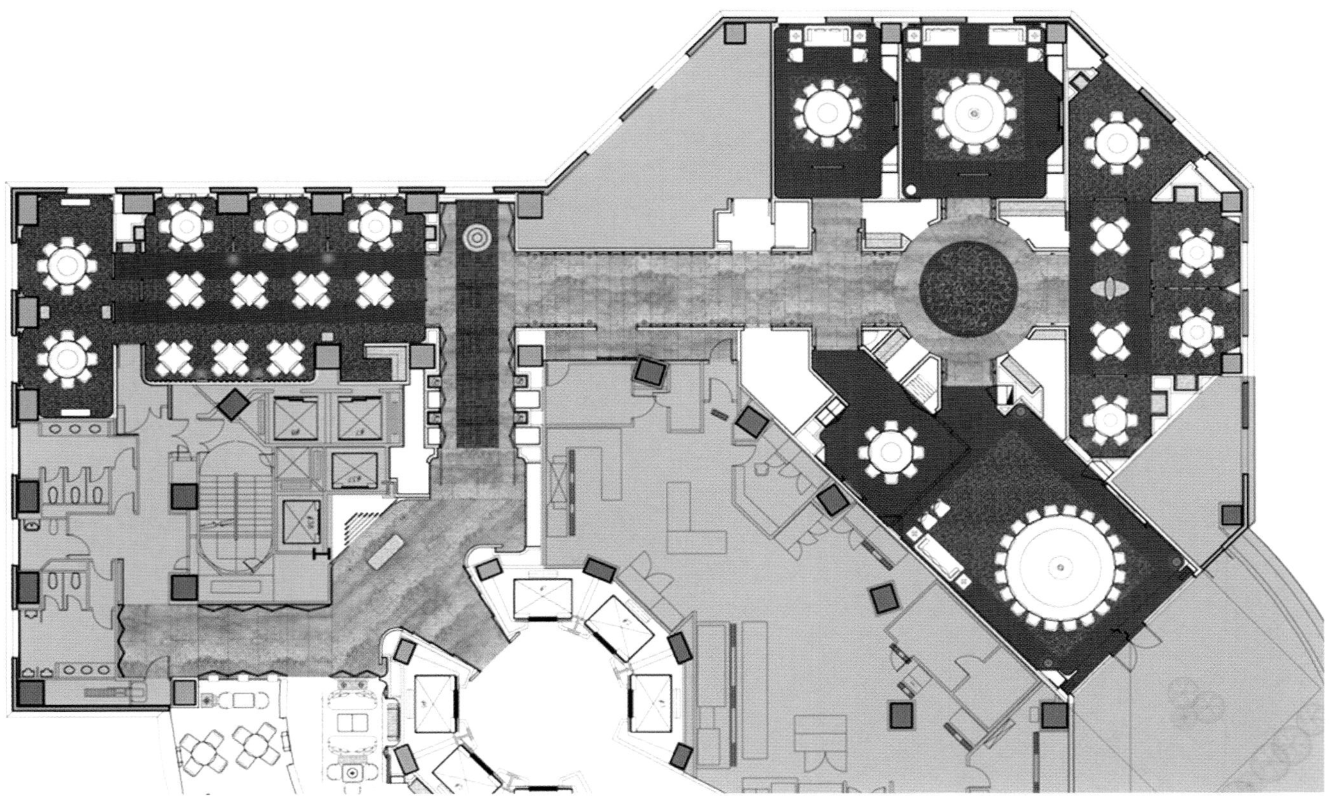

THE FRENCH WINDOW

Designers: Ed Ng, Terence Ngan
Design Company: AB Concept
Location: Ifc Mall, Hong Kong, China
Area: 500 m^2

This high-end French restaurant is a welcoming blend of nostalgia and modernity, where high quality home dining allows guests to eat in style but still let down their guard. The key words when designing The French Window were: upscale, not uptight.

The studio's primary task was to create a distinct impression with the façade, where the identity of the adjacent retailer had been overpowering. In response, it designed iconic, architecturally lit lantern-style columns in wood and wrought iron trim to project out into the mall. Modern lanterns and lush vertical garden panels lead guests part-way along the shared promenade and into the 500-square-meter deep-set restaurant along the mall's sea-facing wall. Yet to arrive there, a long transitional space is necessary; the idea was to echo the journey through the dim hallways of a French chateau, and guests get to encounter the restaurant's impressive wine cellar, the French window motif in dramatic wood louvers and the refined olive palette of cool grays, creams and taupe, before being greeted by the view-drenched lounge and conservatory.

With Victoria Harbour laid out through double-height floor to ceiling windows, the main lounge has been designed to blend rather than compete. The subdued palette creates a sense of tranquility and the scheme is decorative in a subtle, architectural way. Hand-beaten wrought iron and textured glass meet in a series of modern screens, inspired by Parisian antiques and reminiscent of antique French windows.

AB Concept wanted to give the space a relaxed residential feel, and this comes through in the ornate taupe carpeting and the choice of casual, modern Gallic-inspired furniture pieces. The lighting also works to this effect: functional down lights have been discreetly placed and instead

the eye is drawn to decorative floor lamps and suspended lights, all in wrought iron with strong French detailing, and inspired by lucky finds made by the designers in a Parisian antiques market.
This ornate accent tempers the coolness of the more hard-edged materials, like the iron, glass and mosaic floor tiles, as does the careful use of texture. Pearlised paint on the ceiling softens the ambience with a satin glow. Other playful touches among the furniture add characters, such as the copper pan inspired chandelier, small condiment, wine and food display tables and a pantry concealed within a vintage cupboard.
The restaurant is smartly delineated into zones. The "loggia" evokes a patio, with mosaic stone tiling, simple vintage-inspired furniture and oversized lamps. Long banquettes, table clusters, the use of carpet and the wrought iron screens then create a layering effect. The screens are flexibly designed on tracks, and can be slid away to open up the venue for large events.
Cozier U-shaped banquettes comprise the inner dining layer and are elevated to make the most of the view, though anyone facing inwards can instead enjoy the whimsy of ornate trompe l'oeil mirrors on the back wall. Further inside the salon, a discreet VIP room features small naturally French windows connected with the show kitchen so that diners can interact with the chef. Framed photos were handpicked by the designers in Paris to complete the ambience, while a small ornate cabinet houses the chef's personal collection of wineglasses.

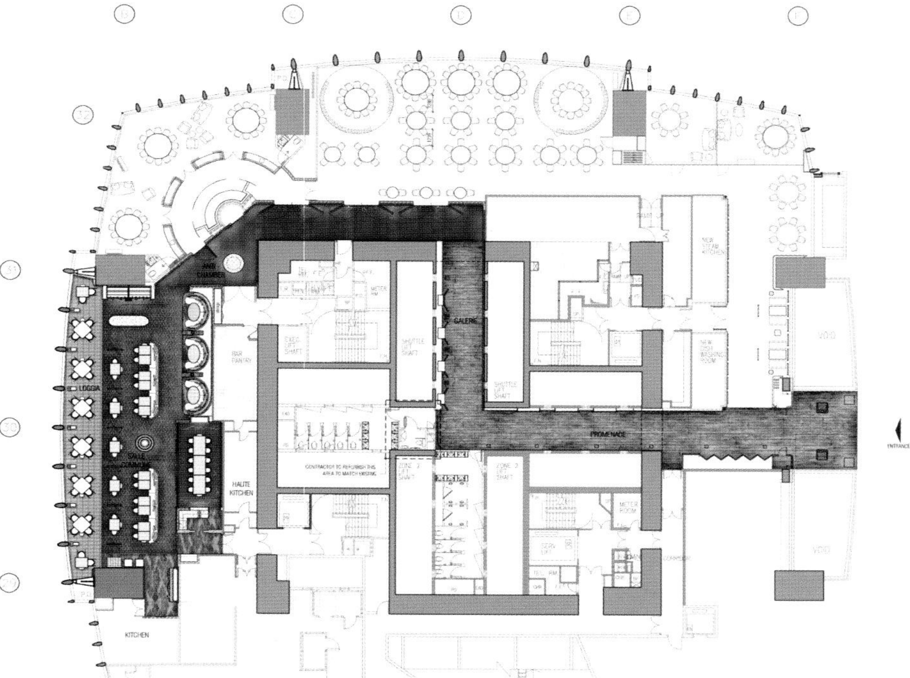
LOGGIA
SALLE COMMUNE
BAR PANTRY
HALITE KITCHEN
KITCHEN
METER RM
EXEC. LIFT SHAFT
SHUTTLE LIFT SHAFT
GALERIE
CONTRACTOR TO REFURBISH THIS AREA TO MATCH EXISTING
ZONE 2 LIFT SHAFT
PROMENADE
NEW STEAM KITCHEN
NEW DISH WASHING ROOM
METER ROOM
VOID
ENTRANCE

这间高级法国餐厅的设计糅合了怀旧与摩登，令人喜出望外。这里优质的传统佳肴能够令客人轻松、自在地享受时尚的美膳。The French Window的设计原则是：高尚而不拘谨。

AB Concept的首要任务，是为餐厅入口的位置设计一个鲜明的形象，因为与其相邻的商店皆出售为人熟悉的高级品牌，且拥有显赫的名声。因此，AB Concept设计了一对极具标志性的发光灯柱，以木材及锻铁制造而成。客人踏进餐厅，首先映入眼帘的便是与其相邻商店共享的修长通道，沿途尽是现代式的灯笼及茂盛的垂直式植物镶板。走到尽头，便是柳暗花明的全海景主厅，面积为500平方米。不过，在到达那里之前，需要经过一段长长的走廊。设计师有见及此，以法国大宅昏暗的长廊为灵感，与餐厅的走廊相映成趣，客人沿途更可参观餐厅庞大的酒库、大型木百叶窗、柔和的褐灰棕榈色调，直到最后才令主厅及观景台的醉人海景映入眼帘。

两层楼高的落地大窗将维多利亚港风景引进室内，因此，餐室主厅的设计原则旨在衬托美景，而非喧宾夺主。餐厅布局采用细腻的橄榄油主色，营造出安静的氛围，装饰主题比较低调的同时，又充满建筑美感：装饰图案遍布整个餐厅，独特且新颖的图案结合手打锻铁及纹饰玻璃，出现于一排排时尚的屏风之中；屏风的设计灵感源自于法式古建筑，外形仿如古典的法式窗户。

AB Concept希望令客人拥有轻松的家居体验，因此，餐厅内铺设了褐灰色的纹饰地毯，配以现代法式家具。灯光的布局亦尽显心思，照明的灯具设置在隐蔽的位置，务求令客人的目光移向餐厅内的坐地灯饰和吊灯，灯饰皆以锻铁装裱，充满法国风情，其灵感源自于设计师在巴黎古董市场淘宝而来的战利品。

华丽的装饰中和了餐厅内其他材料如锻铁、玻璃及马赛克地砖所带来的冰冷之感，对质感的小心处理也有着异曲同工之妙。而别出心裁的家具更是画龙点睛，例如，以不锈钢锅为灵感的吊灯、小巧的调味瓶、美酒佳肴展览专桌，以及隐身于古董橱柜后面的备餐室等。

整个餐厅被精心规划为不同区域。“凉廊”以阳台为蓝本，配以马赛克石砖和简约、仿古的家具，以及特大灯饰。长长的宴会桌、错落有致的餐桌、奢华的地毯，以及锻铁屏风，都令空间层次分明。屏风都安装在轨道上，可灵活推动来扩阔空间。

U型的沙发令客人倍感亲切，构成了餐厅的内围。U型卡座位于高台之上，客人站在此处便可以尽览胜景。而面向内围，亦可透过墙上一列幻景镜子欣赏不一样的风景。

餐厅更深入之处是一个贵宾室，设有小型法式窗户，并与展览厨房相连，客人可尽情与厨师交流。墙壁上挂满由设计师精选的照片，均以巴黎为背景，与餐厅气氛丝丝相扣。此外，贵宾室中还摆放了一个小型饰柜，用来陈列大厨私人珍藏的酒杯系列。

LIBAI LOUNGE, SHANGRI – LA TAIPEI

Designers: Ed Ng, Terence Ngan
Design Company: AB Concept
Area:180 m^2

This upscale bar and lounge were designed with romantic High Tang poet Li Bai in mind, and they blend the sophistication of a members' club with a note of scholarly gravity. Working from the wedge – like shape of the room plan, AB Concept based the design loosely on a traditional Chinese scholar's pavilion and layered it with a soft symmetry. This is achieved architecturally with an arrangement of pinewood panels in layers along the ceiling; distressed by hand for a textured effect and lit from behind, the panels follow the curve of the room and are echoed below by the step and grain of the travertine flooring.

The cool grey palette is brightened and warmed by distressed woods and amber accented textiles while the soft inviting seating serves sturdy bronze table tops. There is a strong residential vibe here: console tables and lacquered screens stand sentry, and the lighting is provided by decorative lamps with oriental design motifs. Though large windows give the lounge an airy disposition during the day, in the evening, its character changes, aided by softer lighting, live music and an innovative display behind the bar that allows the art work to be rotated and replaced by an array of bottles.

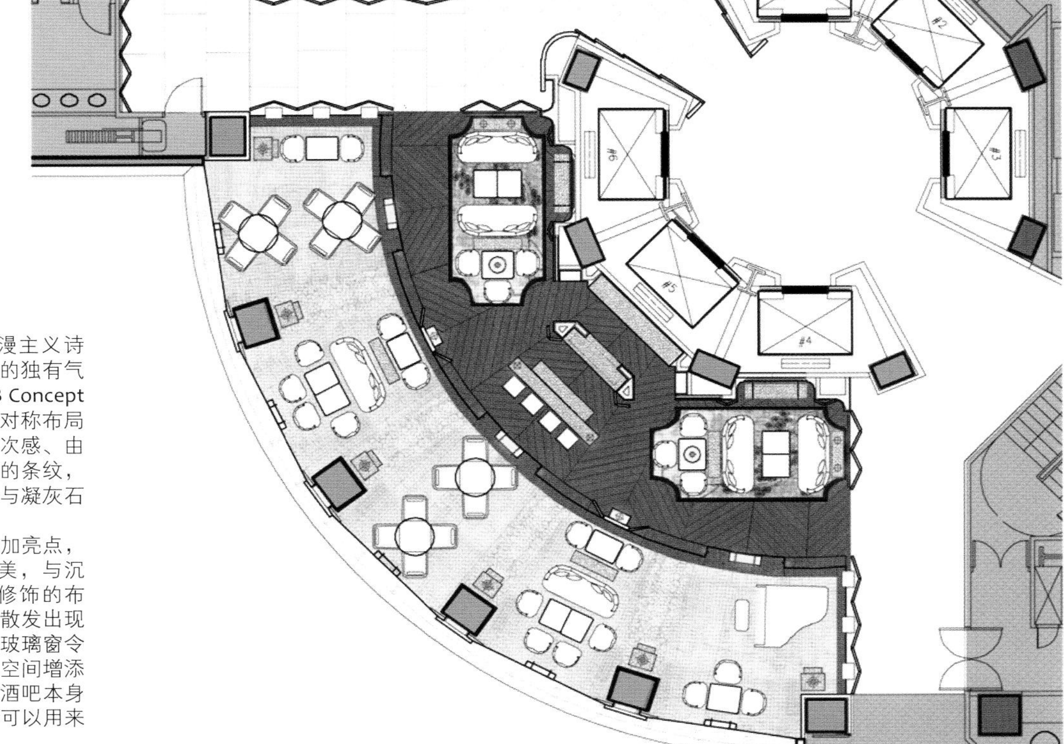

这个高档的酒吧和会客厅的设计从盛唐浪漫主义诗人李白身上汲取灵感，装潢既蕴含着会员俱乐部的独有气派，又散发着学者的稳重睿智。空间呈楔形，AB Concept的设计构思源于中国古典式书房，并采用隐约的对称布局以平衡空间感。这种建筑结构充分展现出富有层次感、由板材镶嵌而成的天花板；面板用铜丝刷出仿风化的条纹，并依照房间的弧度排列，光线从背后透出，正好与凝灰石地板的纹理相互映衬。

仿古木和琥珀色的纺织品为冷色调的灰色添加亮点，增加温暖和光亮感。柔软、舒适的躺椅线条优美，与沉实、时尚的古铜色桌子形成强烈对比。不过于修饰的布置——操作台桌子、亮丽喷漆的屏幕和装饰吊灯散发出现代东方的韵味，犹如置身私人居庭。在日间，大玻璃窗令酒吧餐厅更加明亮通风。入夜后，柔和的灯光为空间增添一份细腻的质感，配合轻松的现场音乐，彰显了酒吧本身特有的个性。旋转式的层架设计既可以储物，亦可以用来展示艺术品。

CLANCY'S FISH PUB

Designer: Paul Burnham Architect Pty Ltd.
Design Company: Paul Burnham Architect Pty Ltd.
Location: Australia
Photographer: Jody D'Arcy Photographer

The plan was to revive and transform the former venue into a new, family friendly place suited to the beach front location. The designers set about to create a casual and lighthearted environment evoking the feeling of a summer holiday shack appropriate for the sun, sand and sea of an Australian summer.
The clients – Joe and John requested color, color and more color. For speed and economy, the new works were limited to the treatment of floors, walls, ceilings and the provisional of all new furniture.
The large space was stripped out and the vast raking ceiling was painted out in a dark charcoal, providing a background to the twelve oversized lavish fabric chandeliers. The new lights are suspended from the ceiling by a playful mass of vibrant fabrics, suggestive of the color and brightness of beach towels. The new chandeliers are a juxtaposition of the traditional and whimsical.
The concrete floor was painted over with children's games, patterns and mermaids. Multi-colored slatted chairs assembled from recycled timber and old street signs added to the 1960's holiday atmosphere. Other lamps and furniture were sourced from second-hand stores.
The spectacular beach setting with its wide, bright, glaring white sand and sparkling blue water provides a complete contrast to the new venue. Clancy's Fish Bar, City Beach is a festival of color creating a bright and frivolous carnival destination.
The make-over was completed in just seven weeks to ensure the new Clancy's to benefit from the remaining Perth summer. The project benefited from the clients enthusiasm for creating something unique and fun. Clancy's is a place where families and children can come in off the beach with sand on their feet, enjoying fish, chips and bottles of beer.

OF COURSE WE
HAVE GREAT DRAUGHT BEERS
FISH
CHIPS.
BEER
WA BEERMASTERS

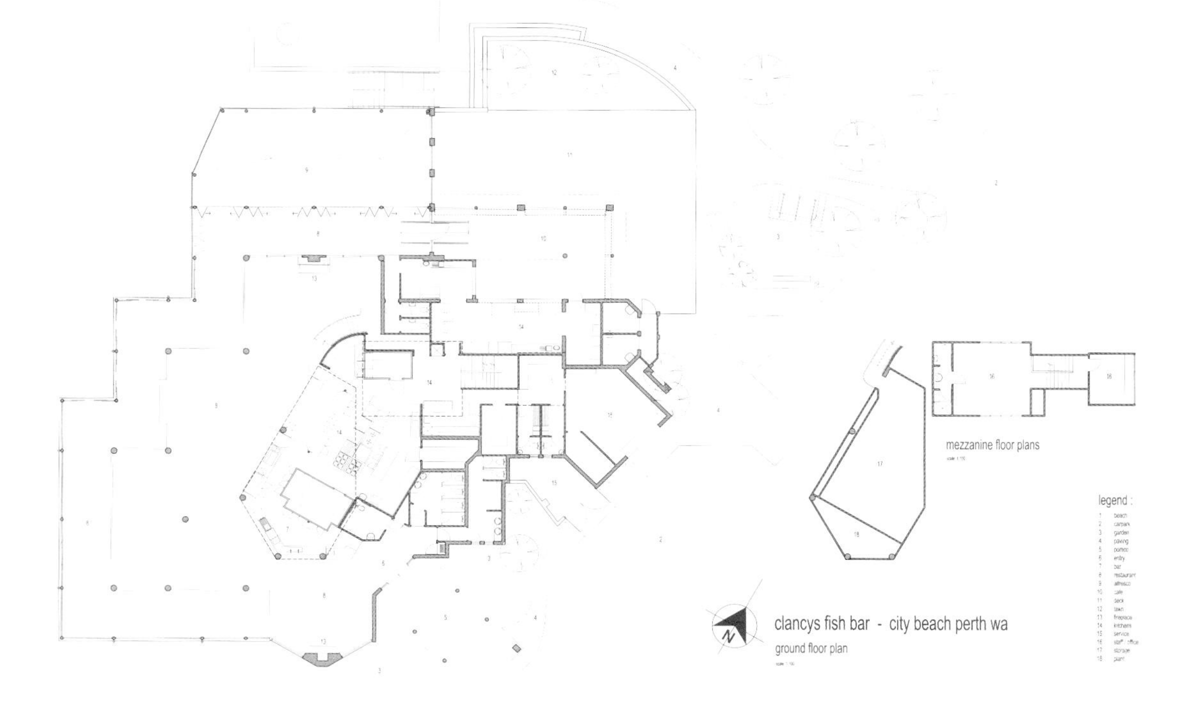

该项目计划将原有场地改造为新的、适合家庭活动的海滨度假胜地。于是，设计师着手营造一个休闲、轻松愉快的环境，为夏季的澳大利亚建造一个有阳光有沙滩的避暑胜地。

色彩是业主的要求重点。由于受时间和财力所限，此次改造只限于地面、墙壁、天花板和所有家具。大空间被分离出来，天花板被绘成黑木炭色，目的是为12个超大型豪华吊灯提供一个背景。而这些集传统与现代风格于一体的新吊灯就悬挂在天花板上。

混凝土的地板上镶嵌着儿童的游戏画和美人鱼的图案。设计师利用回收的木材，自己动手组装了五彩板条椅子，再配上旧街道的标志，让人不禁联想到20世纪60年代的节日气氛。其他灯具和家具则均来自二手商店。

宽广、壮观的海滩上阳光明媚，刺目的白色沙滩和波光粼粼的蓝色海水与原有场地形成鲜明对比。Clancy鱼吧和海滩，营造了一种明快的狂欢节气氛。

该项目在短短7个多星期内完成，如此迅速是为了确保新Clancy鱼吧在短暂的夏日时光中能够充分盈利。Clancy鱼吧现已经成为一个深受父母和孩子喜爱的海滨度假胜地。他们赤足在沙滩上，品尝鲜美的鱼，咀嚼酥脆的薯条，畅饮冰爽的啤酒，尽情享乐。

MY SQUASH CLUB

Designer: Anna Matuszewska – Janik
Location: Poznań, Poland
Photographer: Paweł Penkala

My Squash is a sports club that offers six professional squash courts, but also yoga and dance classes, aerobic, art classes for children, etc.
The club also features a small restaurant and a hotel located nearby.
The place was designed as a typical club that would attract a group of people and convince them to stay not just for one hour of play, but for longer – that is why the investor decided to open the bar and restaurant, for the purpose of having the interiors equipped with huge, comfortable sofas and ensuring the extra activities.
The designer's main goal was to create interiors completely different from the traditional image of a sports club that can be seen around us (in Poland). The new club was to be modern, easy on the eye and surprising.
It was supposed to be stunning and outraging, inspiring and confusing at the same time, and surprising with its creativity.
The interior stands out with its white and shiny epoxy floor and a nearly fanatic consistency in colors.
The design included the installation of fantastic lamps by the internationally recognized Polish Puff-Buff workshop, as well as Cumulus lamps by Daria Burlińska, a young designer.
The designer also used standard Scala sofas (Mebelplast) and Dr. Yes' chairs designed by Starck for Kartell.
The remaining elements of interior design were made to order, according to the design.
It is also important that the investor was open to various ideas and did not limit the investment to a fixed budget, but gave the designer total freedom in their creative work.

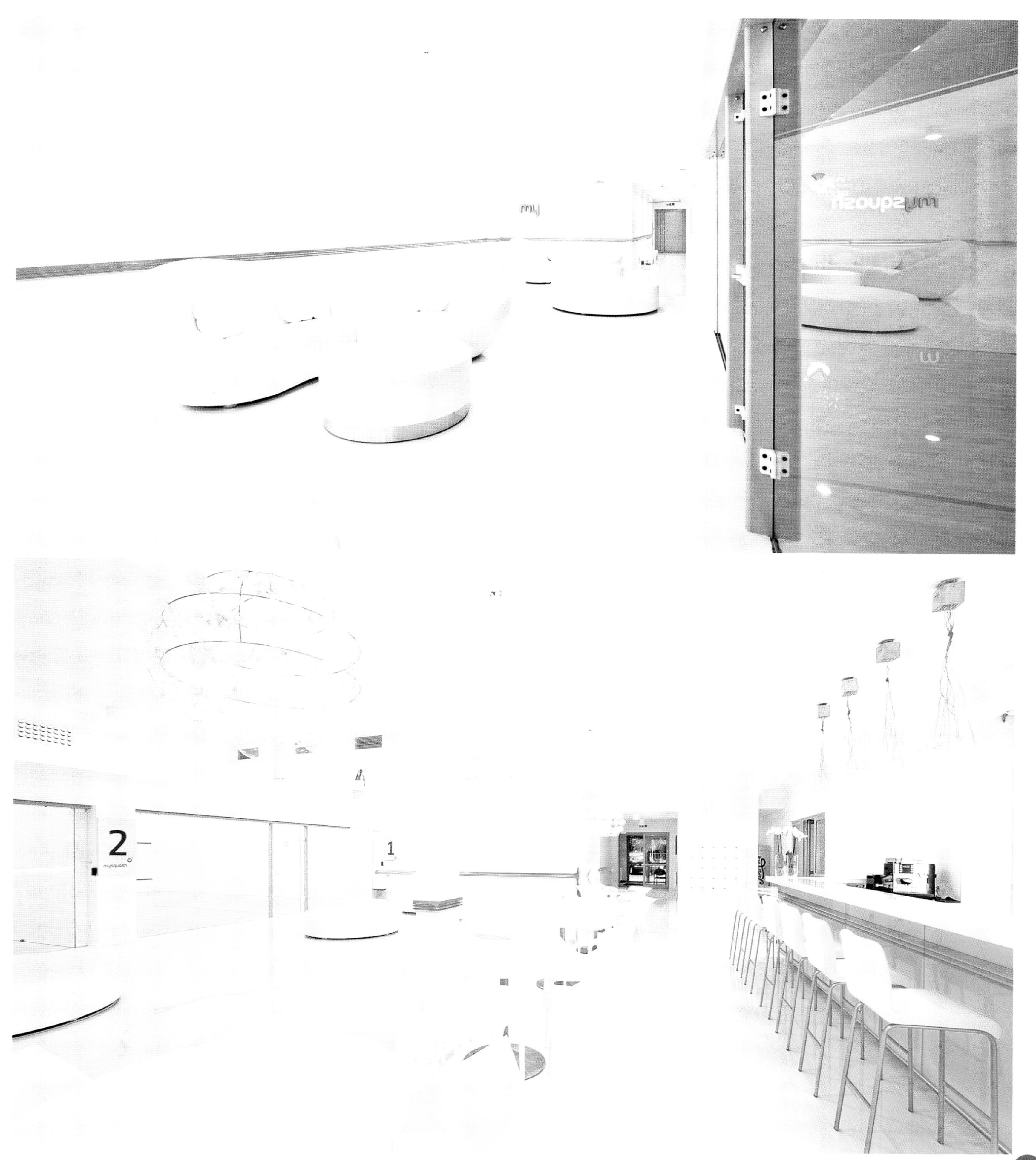
my
2
mysquash
1

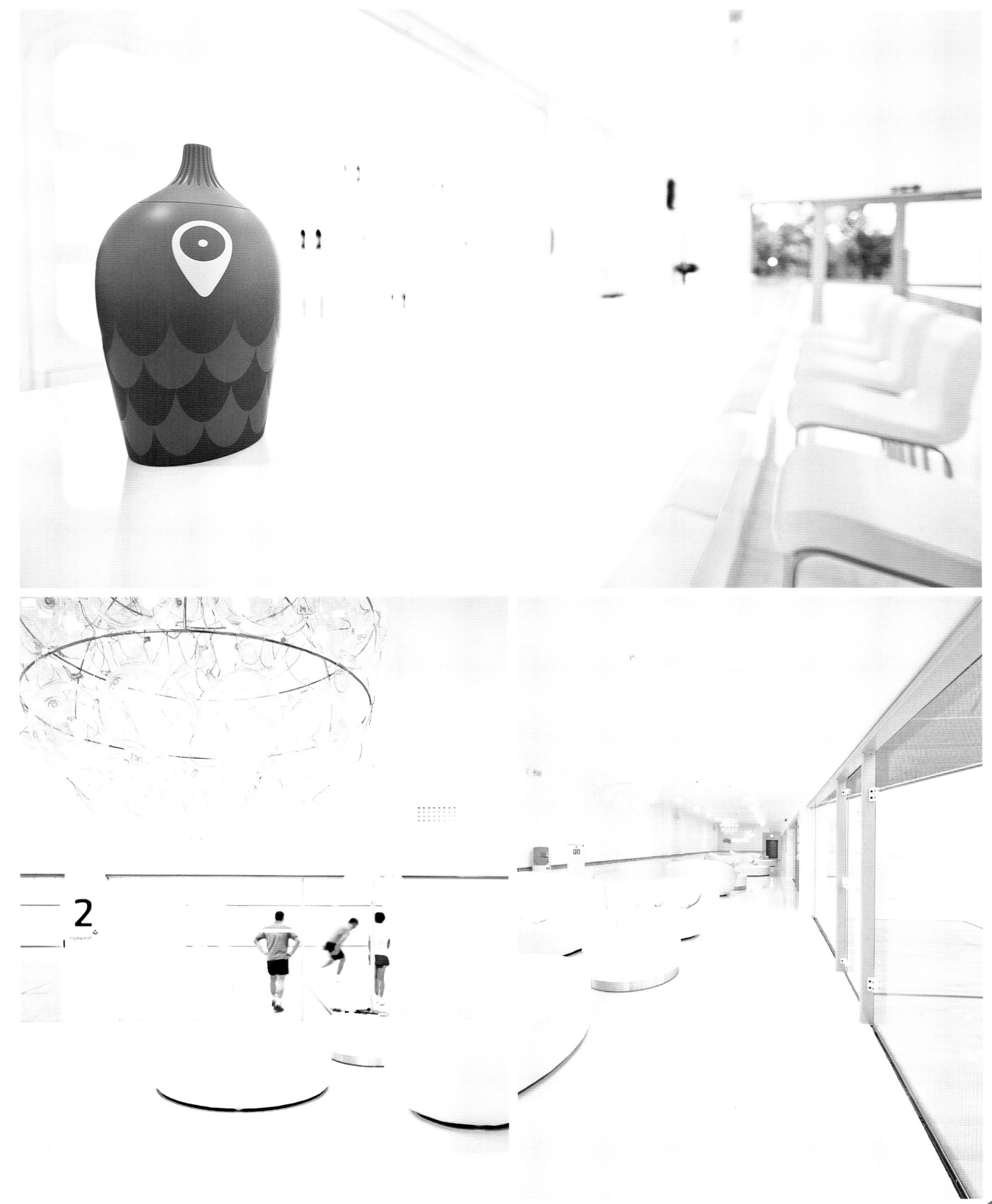
2

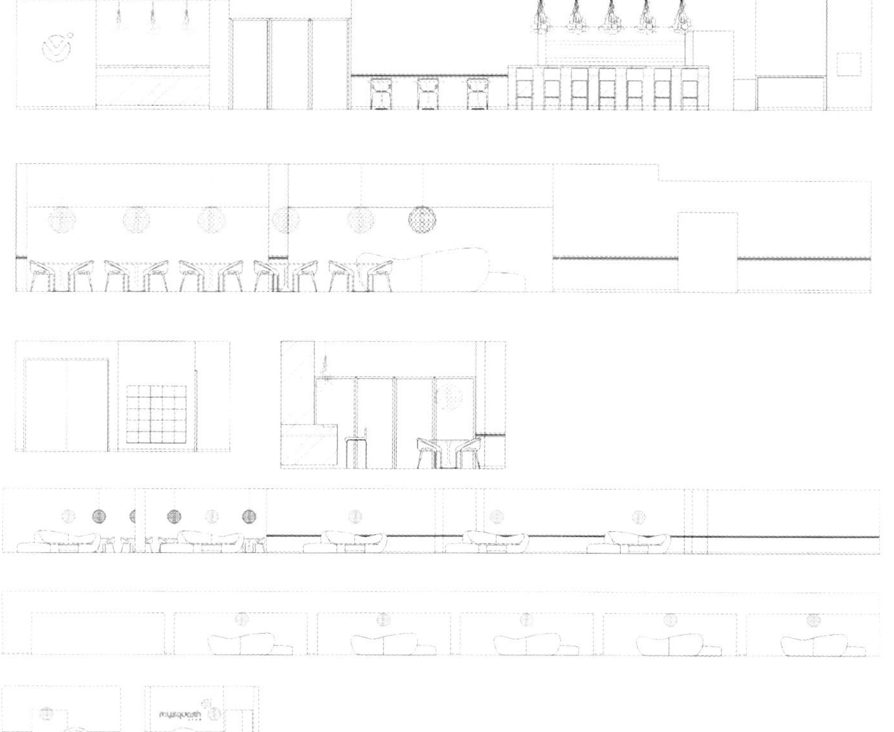

My Squash是一家体育俱乐部，设有6个壁球场，同时也给孩子们提供场地练习瑜伽、舞蹈，做有氧运动或者上艺术课等。

有别于其他俱乐部的是，这里有一个小餐厅，附近还有一家旅店。

如此设计的目的不仅仅是要吸引客人前来健身一个小时，而是还能够让他们长时间停留——这就是投资者决定开设酒吧和餐厅的目的所在；馆内还摆放着舒适的大沙发，这一切保证了在健身之后，客人还可以选择其他活动。

设计师的主要目的是在室内设计上，创造一个与周边常见的传统体育俱乐部完全不同的印象，新俱乐部要富含时代气息，令人耳目一新。

空间令人赞叹的设计让人激情澎湃，充满了诱惑。

内饰亮白的聚氧树脂地板色调匀称，格外醒目。

设计包括安装国际闻名的Puff-Buff台灯、Cumulus台灯，以及年轻设计师Daria 的作品。

设计师也用到了SCALA品牌中有代表性的沙发和Starck for Kartell 公司的Dr. Yes座椅。

其他室内设计元素保持与整体设计风格相一致。

投资者对该设计方案很认可，非但没有限制投资底线，反而为设计师提供了一个自由创造的空间。

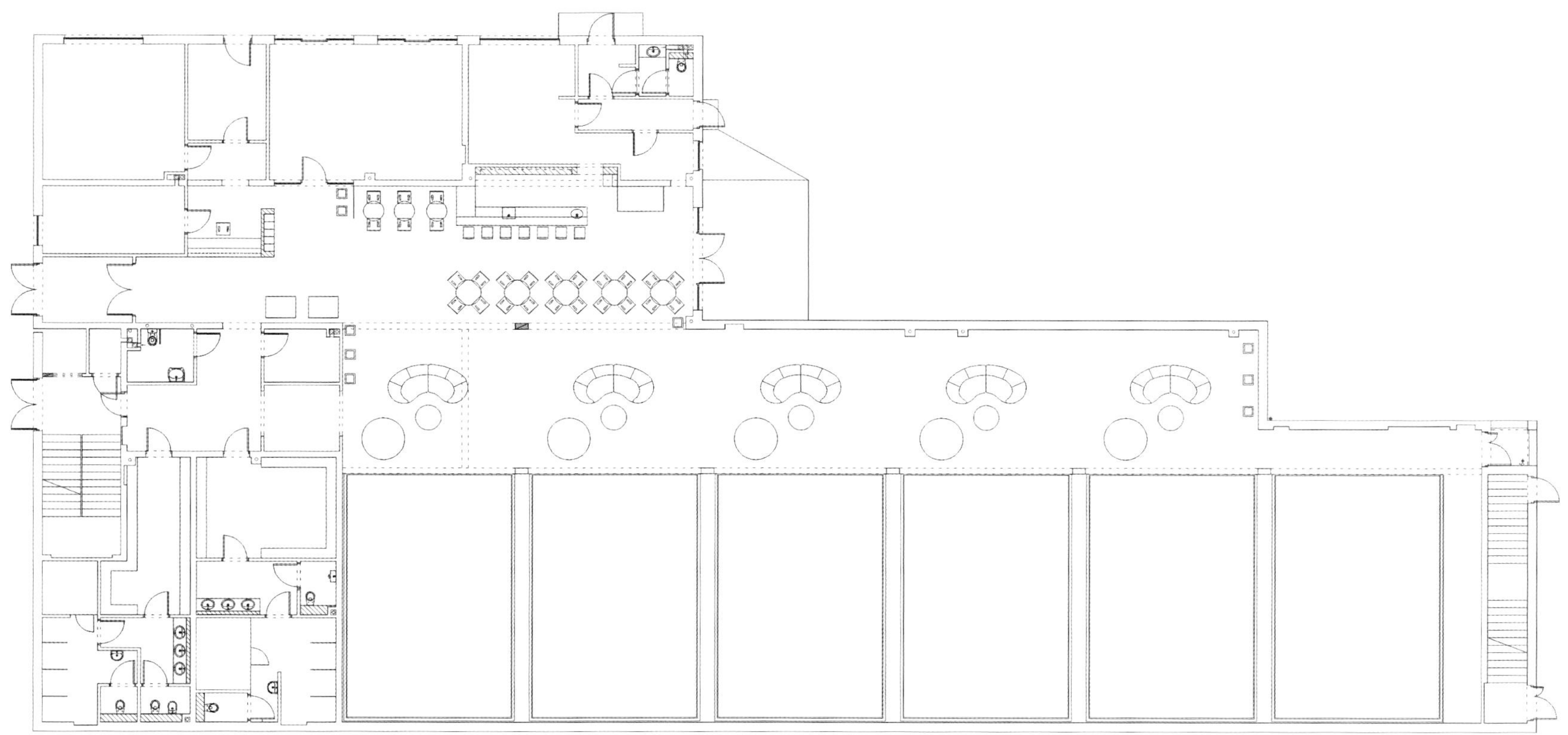

WIENERWALD

Designer: Ippolito Fleitz Group - Identity Architects
Design Company: Ippolito Fleitz Group - Identity ArchitectsLocation: Munich, Germany
Photographer: Zooey Braun

The space has been organised to ensure good visitor guidance, crucial in a self-service restaurant, as well as respecting the need for a differentiated selection of seating. Upon entering the restaurant, the guest is guided towards a frontally positioned counter, which presents itself as a clearly structured, monolithic unit. Menu boards suspended above the counter visualise the range of food on offer. The food itself is also visible. The wall is covered in anthracite mosaic stones, into which frameless, stainless steel units have been precisely inserted, thereby underscoring the high standard of the products.

Order and payment terminals occupy the far ends of the white, solid surface counter. The chopping station is in the middle. After ordering, this is where salads are chopped, and chicken is portioned and toppings are added from containers set into the counter under the guests' watchful eyes. In the wall adjacent to the payment terminal, a display refrigerator stocks drinks and desserts.

In front of the service counter is a service station made of white solid surface. It stands on golden chicken legs and looks expectantly towards the entrance. Green instructions and Wienerwald chickens set into the rustic wood floor show the guest how to navigate the ordering process.

The dining area offers a range of seating options catering toward different requirements. White solid surface high bar tables are available for guests with little time on their hands. These are supported by a single leg with a tapering cylinder at its foot, recalling the traditional turned table leg. Alternative seating is available in an elongated seating group upholstered in brown, artificial leather, a reflection of the traditional Wienerwald seating niches. Overlapping, rough-sawn oak panels on the rear wall quote the forest theme. Round mirrors printed with the outlines of tree and forest motifs are set into this wall. Different-sized pendant luminaires at varying heights hang over the tables. These are sheathed in a roughly woven fabric in three shades of green and ensure a pleasant atmosphere. Forest images occupy one side wall, as well as transparencies on the windows. The view into the restaurant from the outside thus becomes a multi-faceted experience in which the individual elements on the mirror and glass surfaces reflect and overlap one another, making the brand world a truly holistic experience.

SALATE
HENDL
HENDL
MENÜ
EXTRAS

0
1
2
3
5m

该空间的设计旨在为客人提供一个好的路线指导，这对于一家自助餐厅来说至关重要。进入这间餐厅，客人就被引向前面的吧台，这是一个独立的空间。吧台上悬挂着菜单板，上面展示着本店的菜品图片。

买卖实物是在白色的吧台上进行的。食物中转站设在餐厅中间，在点餐之后，厨师在这里拌沙拉、烹制鸡肉，并添加配料，这一切都在客人的眼皮底下完成。在餐厅的角落里，摆放着一个放满了饮料和甜点的冰柜。

在服务台前面有一个白色台面的服务站。服务站的两根台柱被设计成了金色的鸡腿，爪尖朝向餐厅入口，给人一种期待进入之感。

入口处的绿色指示牌以及地板上的Wienerwald鸡图案，引领客人选择食物。

餐饮区根据要求提供不同的座位以及食物。白色的高吧台方便客人享受美食。桌子都是单腿支撑的，让人想起传统的单腿旋转椅。

等餐客人的座位是一个棕色的皮质拉长座椅，座椅后面的木质墙壁凸显了餐厅的绿色森林主题。印有树木和森林图案的圆形镜子挂在墙上。不同高度、不同大小的吊灯悬挂在餐桌上方。树形装饰布满了整个面墙，同时也映射在玻璃窗上。从外面看餐厅内部，会有一种多方位体验的感觉。镜子中和玻璃上的映像相互呼应、重叠，给人一种真正的全方位体验之感。

Heute bleibt die
Küche kalt,
wir gehen in den
Wienerwald.
Wiener
wa

MASH

Designers: Ippolito Fleitz Group - Identity Architects
Design Company: Ippolito Fleitz Group - Identity Architects
Location: Stuttgart, Germany
Area: 500 m^2
Photographer: Zooey Braun

The Mash brewery, a popular party destination, was seeking to reposition itself through comprehensive refurbishment and the introduction of a mixed concept. The new Mash is designed to cater for events as a club, restaurant and all-day bar. The large bar, which is stunningly illuminated by more than 2,500 plastic batons, takes center stage in the space. It divides the room into three zones in which two niches enclosed behind silk curtains display a more intimate character. The individual areas are connected by continuous mauve-colored flooring as well as a superimposed ceiling of white printed tiles with a surrealist collage.

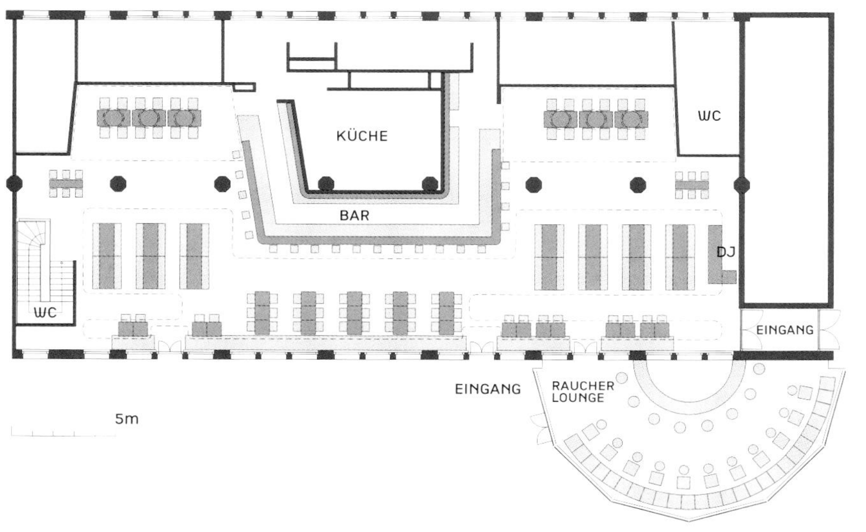

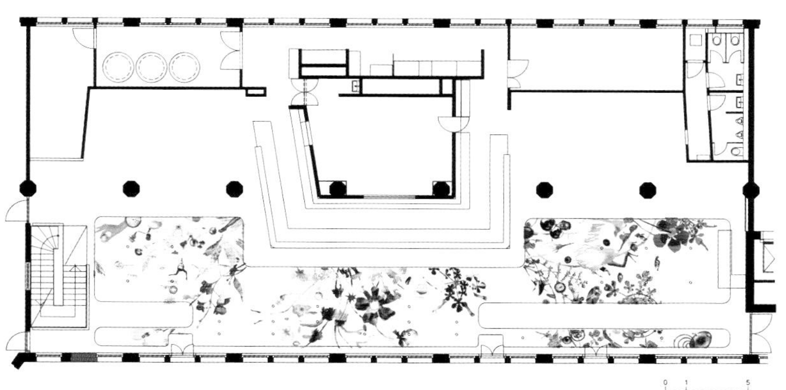

Mash啤酒厂，一个聚会的理想去处，设计师试图通过翻新和引进多元概念来重新定位Mash啤酒厂。新厂的设计有如下功能：俱乐部、餐厅和24小时酒吧。大酒吧在2 500个塑料灯管的照射下，轮廓绚丽夺目。它把这块空间分割成三部分，其中，两个壁龛在丝质帘幕的衬托下尤显隐秘。紫色的地板和白色天花板上的超现实主义拼贴画浑然一体，构成了相对独立的空间。

THE NIGHT MARKET

Designer: Alexi Robinson Interiors and Michael Young
Design Company: Alexi Robinson Interiors
and Michael Young
Location: Hong Kong, China

The Night Market is a Taiwanese restaurant in Central, Hong Kong that serves famous street foods, home-style dishes and renowned Taiwanese drinks. It's a restaurant that personifies the creators' experience of growing up in Taiwan as well as having traveled and lived around the world.

Collaboration between interior designer Alexi Robinson and industrial designer Michael Young, the Night Market brings together a balanced environment through each detail. Alexi Robinsons says, "Designing the interior of Night Market meant capturing the essence and honesty of a Taiwanese market place and presenting it in a relevant and enticing manner to the contemporary diner. Here, I had the opportunity to collaborate with Michael Young and together we drew inspiration from the many characteristic aspects of the markets themselves to create two individual spaces that could cater to varying moods, tastes and party numbers.

"For the 7th floor interior, I wanted to capture the energy, vibrancy and transient "pull up a stool" nature of the outdoor market which meant the folding elements, weathered materials, hard/durable surfaces and a light installation reminiscent of the abundance of illuminated signage present at night.

"By contrast, the 6th floor feels as if you are treated to the living quarters of one of the many quaint wooden shops that usually complement the night market; an intimate space of warm materials and screened off windows for a more internalised perspective, delicate handmade materials and precious objects tells the story of finding beauty in the otherwise broken or discarded; the very details often encounter within the foundations of Taiwan's night market culture.

“夜市”是位于香港中环的一家台湾餐厅，提供著名的街边小吃、私房小菜和台湾饮料。创建者在台湾的成长经历和旅居世界各地的经验赋予了该餐厅人性化的特色。

室内设计师Alexi Robinson与工业设计师Michael Young合作，通过餐厅的每一个细节，营造出一种平衡的环境。Alexi Robinson说：“进行夜市餐厅的内部设计意味着要捕获台湾餐厅的精髓与真诚，并将餐厅以恰当的、迷人的方式呈现给当代的食客。在这里，我有机会与Michael Young合作，一起从众多餐厅的特色设计上获得灵感，打造出两个独立的空间，以迎合不同人的心境、品位和就餐需求。”

“在七楼的内部设计中，我想体现的是外面市场的活力、随意和方便，例如，拉过来一个凳子就可以就餐。所以在设计中，运用了折叠元素、风化料、坚硬且耐久的表面和灯的装置，让人联想起夜晚灯光招牌的炫目多彩。”

“相比较而言，六楼的内部设计使人感觉身在古怪有趣的木质结构商店中，这一点往往是夜市的一大特色。暖色调的材料和无遮挡的窗户打造出一个舒适而隐秘的空间，精致的手工材料和珍稀物件仿佛在讲述着在破碎或者遗弃中寻找美的故事，每一个细节都基于台湾的夜市文化。”

ALICE OF MAGIC WORLD

Designers: Katsunori Suzuki, Eiichi Maruyama
Design Company: Fantastic Design Works
Location: Shinjuku, Tokyo, Japan
Area: 228.11 m^2
Photographer: Diamond Dining

Alice of magic world restaurant's concept is "multi-scene fantasy".
It is fancy and cute which constructed various wonder stories' scenes and all area is different design.
At the entrance, a big book created the wall, making guests feel as if they are small.
In the interior, subsequent to green box seats the motif of the trump kingdom. And the secret party room and trump room constructed the black and red color and trump table, and the trump chandeliers are funny designs.
The main dining with motif of "Madhatter's teaparty" is a wonder and surprise area. All the chairs and tables are featured by different designs. And chandelier stands and tea cups, dishes are sailing like a magic. "heart dining" is big in the shape of a heart table and in the shape of a heart piece's chandelier, and the big mold frame seat with pink color and fancy is popular among girls. At the big molding frame seat, guests can see a gimmick with chessman and flower, bird ranging several times over in the frame art.
All the areas change the design theme, and also some gimmicks, surprise and happiness.
And it changes from a different angle and makes guests enjoy again and again. This shows the true worth of fantasy restaurant.

魔法の国のアリス

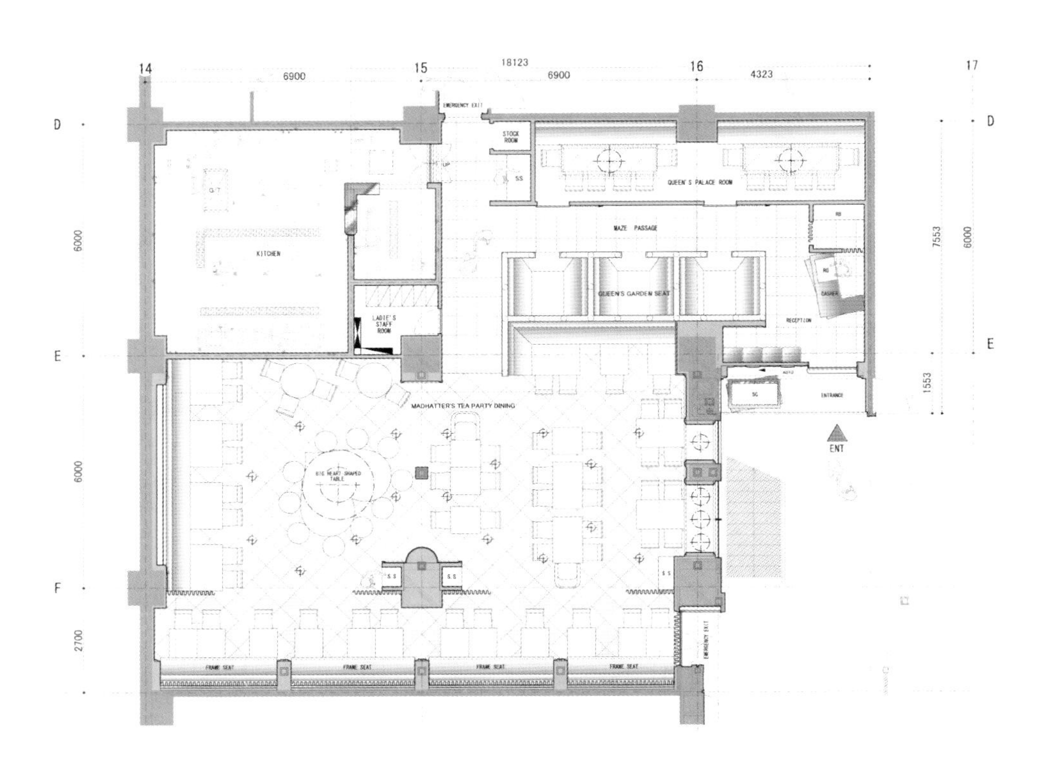
14
15
16
17
18123
6900
6900
4323
D
E
F
6000
6000
2700
7553
6000
1553
EMERGENCY EXIT
STOCK ROOM
S.S
QUEEN'S PALACE ROOM
KITCHEN
MAZE PASSAGE
QUEEN'S GARDEN SEAT
CASHER
RECEPTION
LADIE'S STAFF ROOM
ENTRANCE
ENT
MADHATTER'S TEA PARTY DINING
BIG HEART SHAPED TABLE
S.S
S.S
S.S
FRAME SEAT
FRAME SEAT
FRAME SEAT
FRAME SEAT
EMERGENCY EXIT

奇景爱丽丝餐厅的设计理念是集多个梦幻场景于一体。

餐厅中建立各种不同的梦幻场景，看起来既梦幻又可爱。

入口处是一面书墙，令客人觉得自身是如此的渺小。

内部的绿色包厢式座椅，给人一种帝王般的感觉。“秘密党派房”、“王牌房”的主色调是黑色和红色，内部的皇室桌椅和皇室吊灯的设计都别具一格。

主餐厅的主题是Madhatter的茶会，这是一个制造奇迹和惊喜的地方。所有桌椅的设计都别具匠心。吊灯、茶具、餐碟都好像具有魔力。“心形餐厅”包括一个心形的大餐桌，天花板上悬挂着一个心形片状物做成的吊灯，角落里摆放着深受女孩子喜爱的粉色大模架座椅。坐在椅子上，客人可以看到鲜花、丛林。

所有的空间设计都在不断变换着主题，为客人带来一串串惊喜和快乐，也不乏时下流行的噱头。

每一个变化都有不同的角度，这一切使得客人惊喜连连，展现了梦幻餐厅的真正魅力。

EL CHARRO

Designer: Cheremserrano Arquitectos
Location: La Condesa, Mexico
Area: 200 m^2

This Mexican restaurant located in La Condesa, Mexico, allowed the designers to make an intervention in an existing place, taking care of the functional requirements while satisfying and expressing the client's character and personality. This project was done in collaboration with a Mexican artist called Jeronimo Hagerman who designed the ceiling and called it: "Ando volando bajo" ("I am flying low"). The ceiling is covered by a black perforated shield. Purple and pink dissected flowers are hung in the ceiling creating different forms, which evoke Mexican landscapes of clouds and flowers. The moment the client goes inside the restaurant, the smell of wood and flowers fill his lungs. The floor is made of dark stone and carries a wood platform that takes the client into the restaurant. The columns are in the middle covered by wood panels and are lighten stylishly by indirect light. The tables and chairs were designed especially for this project, emphasizing the designers' concern to use furnishing as an element that defines space and not as a mere decorative object. The goal of this project is the reinterpretation of a Mexican icon"El Charro", to create a modern, elegant restaurant with a unique character that expresses its relationship with the beautiful country, Mexico.

这家墨西哥餐厅位于墨西哥的La Condesa城，设计师在餐厅现有的位置上开始进行设计，在满足并体现出客人性格和个性的同时，也要兼顾餐厅的功能性。这个项目是和墨西哥艺术家Jeronimo Hagerman一起合作完成的，他设计了天花板，并把它称作“我在低空飞行”。天花板由一个黑色的多孔挡板覆盖。天花板中央挂满了紫色和粉色的花朵，打造出不同的形状，让人联想起墨西哥的云和花的风景画。一进入餐厅，木头和花的香味立刻沁人心脾。地板由黑色的石头构成，上面有一块木质平台，引领客人进入餐厅。柱子位于餐厅的中央，由木质嵌板包裹着，间接光源很巧妙地将其照亮。桌子和椅子也是为这个项目特别定制的，突出了设计师的创意：将家具作为定义空间的元素，而不仅仅是装饰的物件。该项目的目标是要重新诠释墨西哥的塑像“El Charro”，创建一个现代、雅致的餐厅，具有独一无二的特点，能够表达出它和美丽的墨西哥之间的关系。

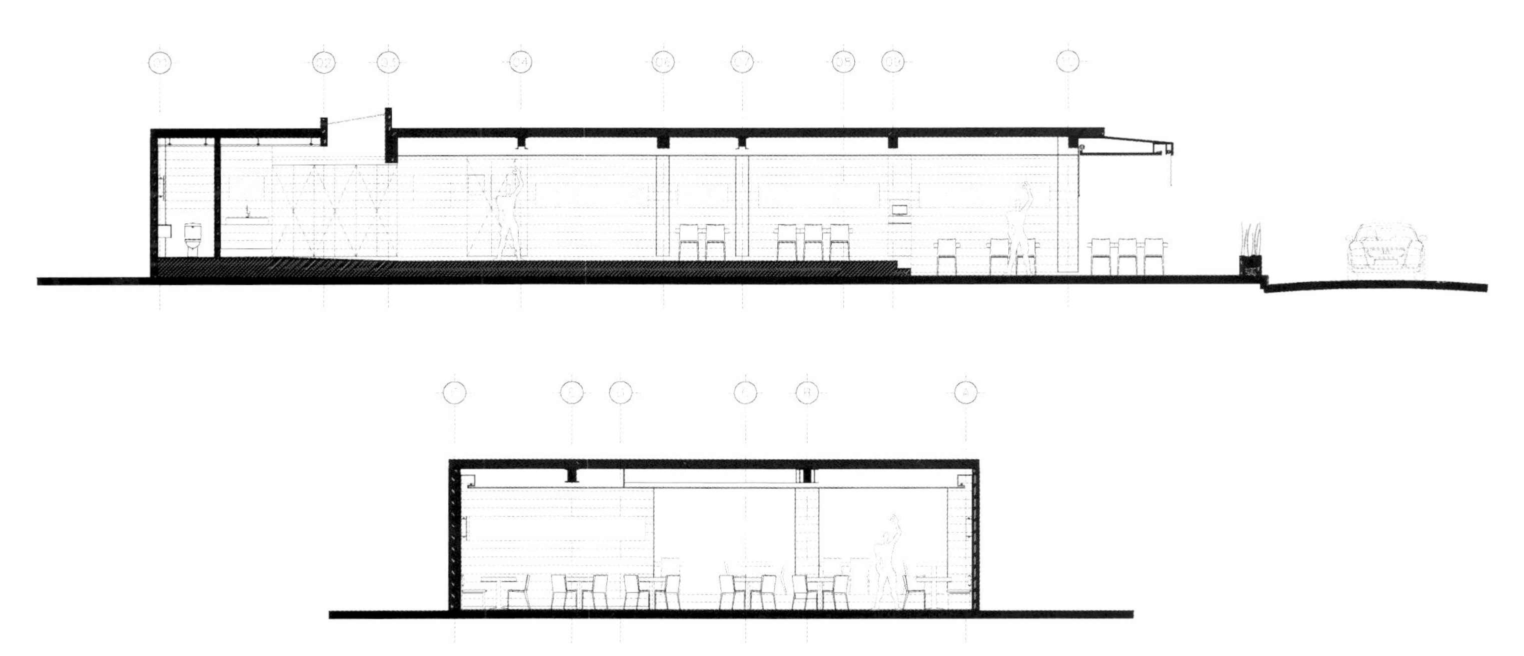

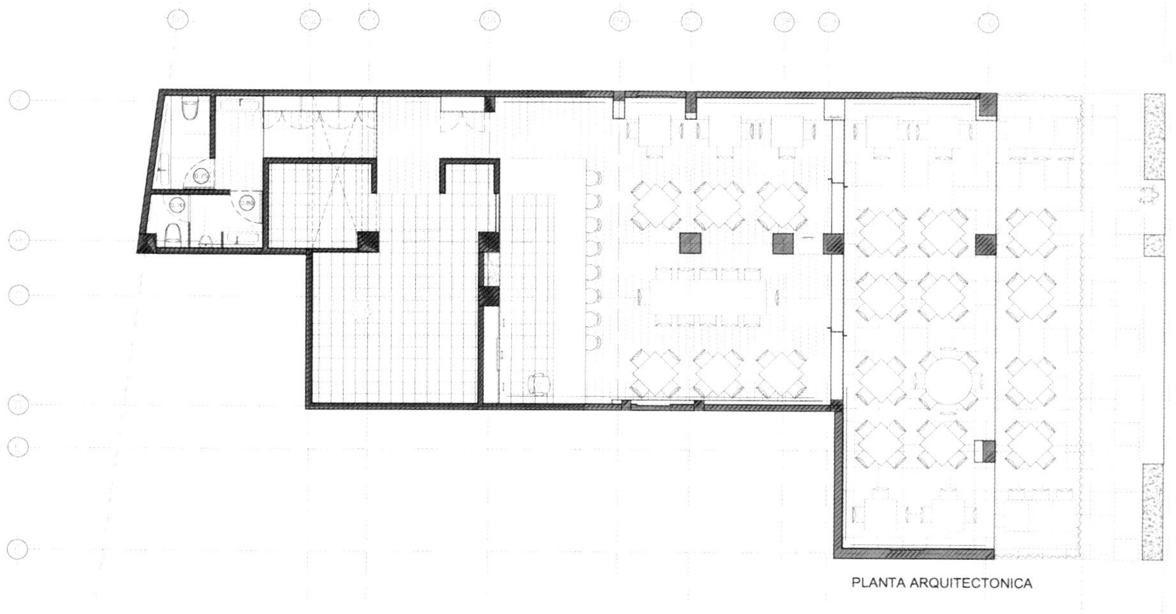

PLANTA ARQUITECTONICA

Design Company: SAQ
Location: Brussels, Belgium
Area: 250 m^2

Located in the center of Brussels, leading up to the Art-Mount, the restaurant KWINT serves as a central meeting spot for the newly-opened "Brussels Square Conference Center". The renovation of the formerly abandoned top-arcades in which the restaurant took its premises is respectful of the architectural heritage. Bringing on a new breeze in this ensemble, SAQ's intervention is minimal: the particular atmosphere is created with a limited amount of elements.

The length of the space is emphasized by both the sculpture hovering over the dining tables and the padded side-wall.

This thirty-meter-long sculpture is a creation of Arne Quinze and erupts from the bar at the end of the room almost like a living and articulated organism. Floating above, the installation gives the seated customers a sense of protection and intensifies the notion of gathering.

The upholstered wall not only functions a s a perfect acoustic absorbent for the ambient conversations, it houses all the essential technical elements and infrastructure: the heating and ventilation system, the electrical installations, but also the annex rooms, such as the kitchen, the toilets, and the lounge.

The wall is garnished with an irregular pattern of dots as if formed by the shadow of the moving sculpture. One color prevailing, the glossy copper presents on the sculpture's skin as well as on the crackled surface of the side-wall openings reacts magnificently with the incoming daylight offered by the generous large French windows.

On summer days, the accordion windows are opened up, making the view over the Brussels ancient center below an intrinsic component of the scenography.

Kwint餐厅坐落在比利时布鲁塞尔市的市中心，一直延伸到Art-Mount。Kwint餐厅已然成为邻近新开的“布鲁塞尔会议中心”参会人员的会后休闲地。它在原已废弃的顶级商场的基础上进行翻新，内部的设计充分体现了设计师对于建筑遗产的尊重。原本的自然风景再加上SAQ设计公司的改造，令此处给人一种焕然一新的感觉。

放置在餐桌周边的雕塑以及填充的隔离墙，使餐厅空间变得狭长。

30米高的雕像位于餐厅的角落，是出自名家Arne Quinze之手，雕像看起来活灵活现，栩栩如生。这样的设计令客人有一种安全感，并且增加了餐厅员工的凝聚力。

隔断墙不仅可以用来隔离周围客人的说话声音，还便于在此安装所有的屋内设施，例如，加热和通风系统、电气装置。与此同时，隔断墙也分割出了一些附带空间，例如，厨房、卫生间以及休息室。

墙上饰有由圆点组成的不规则图案，仿佛是一个由阴影形成的动态雕塑。雕塑通体上下只运用了明亮的黄铜色，开在影壁墙上的门框表面同样采用了铜黄色的裂痕纹饰。充足的阳光透过大落地窗投射进餐厅。

夏季，客人坐在窗边，打开窗户就可以饱览古城布鲁塞尔的美景。

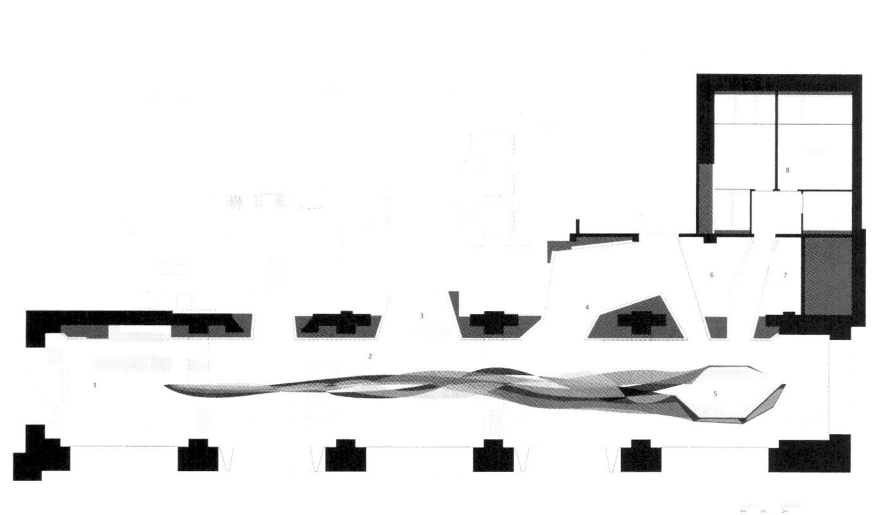

1. Main Entrance
2. Central Room
3. Wine Bar
4. Lounge
5. Bar
6. Storage
7. Technical Room
8. Toilets

PRAQ AMERSFOORT

Designers: Frank Tjepkema, Janneke Hooymans, Tina Stieger,
Leonie Janssen, Bertrand Gravier, Camille Cortet
Design Company: Tjep
Location: Amersfoort, Dutch
Special Thanks: Brandwacht en Meijer & Kloosterboer

In October 2003, two wonderful entrepreneurs came with their wish to start a new unique restaurant concept that would welcome parents and their children without looking like a playground.
Indeed, children don't necessarily want to spend the evening with Donald Duck when their parents go out. The first Praq restaurant opened its doors shortly after and became quite successful and well-known. After five successful years, the two clients realized that the formula had the potential to become a restaurant chain, thus opening two new venues in 2008. Presented here is the first of the two that opened in Amersfoort, Dutch. Again, all have tried to create a restaurant that is inspiring and fun for children, while comfortable and enjoyable for their parents. This in an integrated rather than separated manner, which is more usually the case. The space is characterized by a monumental farm style roof composed of huge massive wooden beams. Within this space, a playful world has been tried to creadted by placing furniture and elements that form crossovers between objects and furniture or furniture that refers to the space itself.
For example, a table becomes a window, a bus or a kitchen. The six-meter-high construction in the center looks like an abstract cloud evoking something of a colorful game while contrasting nicely with the handcrafted architecture. One space is reserved for children and their parents while the other is reserved for adults.
The adult space clearly fits the same world. Only here, the accent is more geared towards cosiness and style. By the way, Praq means mashed food for children in Dutch, however, the plates sereved at Praq are very refined and tasty.

2003年10月，两家优秀企业希望通过合作，建造一个新颖、独特的餐厅，即建造一个家长和孩子都愿意来的餐厅。

事实上，当父母出去的时候，孩子并不希望整晚与唐老鸭待在一起。第一家Praq餐厅开业后不久，就相当成功并且众所周知。经过5年的成功运营，两家企业意识到运用该模式可以建立连锁餐厅，因此，他们在2008年开设了两家新店。这里介绍的是最早在阿默斯福特以及荷兰营业的两家餐厅。

要想开办一家受孩子欢迎的同时也令父母觉得舒服的餐厅，就需要将两种理念有机结合。餐厅的特点是带有一个由巨大木梁撑起的农场式屋顶。在餐厅内部，通过家具摆放的位置，营造惬意的空间氛围。

例如，在窗边放置一张桌子，使16米高的中心建筑看起来像一个丰富多彩的云朵，与手工制作的架构形成了很好的对比。一个空间留给孩子和他们的父母，而另一个则是专门为大人保留的。

大人的专属空间风格一致，此处的设计着眼于舒适和充满格调。与此同时，Praq餐厅精心为孩子烹调食物，在这里出品的每一道菜都精致、美味。

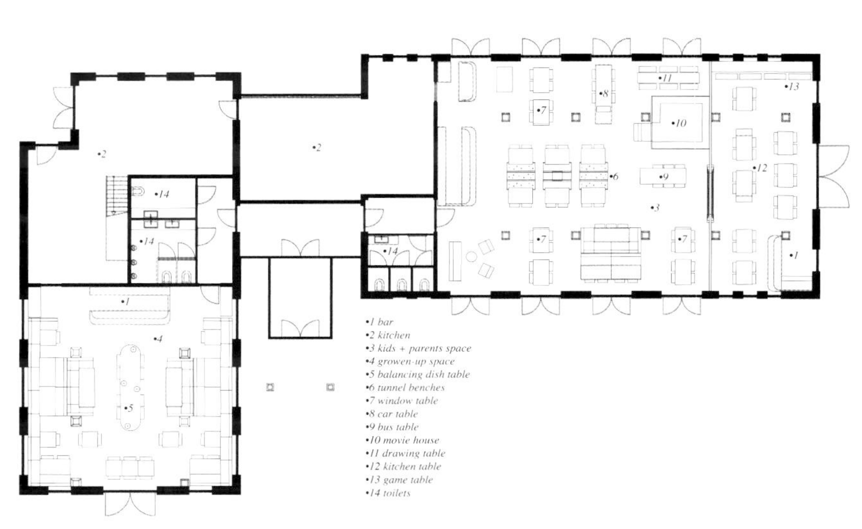
•1 bar
•2 kitchen
•3 kids + parents space
•4 growen-up space
•5 balancing dish table
•6 tunnel benches
•7 window table
•8 car table
•9 bus table
•10 movie house
•11 drawing table
•12 kitchen table
•13 game table
•14 toilets

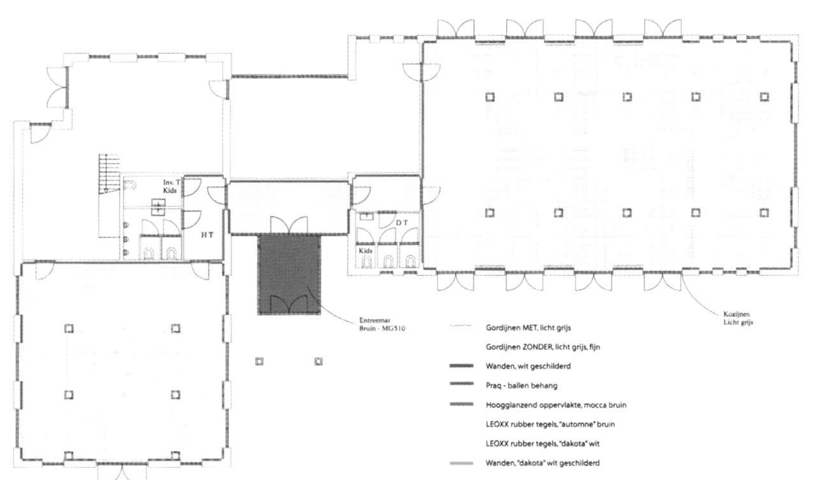
Inv. T
Kids
H T
D T
Kids
Entreemat
Bruin - MG510
Kozijnen
Licht grijs
Gordijnen MET, licht grijs
Gordijnen ZONDER, licht grijs, fijn
Wanden, wit geschilderd
Praq - ballen behang
Hoogglanzend oppervlakte, mocca bruin
LEOXX rubber tegels, "automne" bruin
LEOXX rubber tegels, "dakota" wit
Wanden, "dakota" wit geschilderd

LA NONNA

Designer: Cheremserrano Arquitectos
Location: La Condesa, Italy
Area: 200 m^2

La Nonna is an Italian restaurant located in La Condesa. This project was made in collaboration with DMG Architects. The design premise was to liberate the plan so the restaurant could take the most advantage of the 200-square-meter site. The restaurant fuses its environment with simplicity. Incorporating the use of local materials, the floor is made of dark stone and is built unto the sidewalk, while the walls and ceiling are surrounded by red brick with special cutting on top of mirrors. The mirror was placed in order to enlarge the space and create a game of light and shadows.

The furniture is elegant and simple. At the center is placed a bar with a pizza stove and a wood counter to sit and enjoy a nice wine. The services and the kitchen are placed in a second level connected with the ground floor by an elegant illuminated staircase. While going down to the main level, the designers created a playful mezzanine that catches the eye of the visitor by letting light in and setting a swinging chair, some books and a floor lamp.

The ceiling in the stair and main circulations is painted in black in order to emphasize the design of the red bricks placed in the restaurant area. The illumination is hidden within the bricks and creates a nice rhythm of light that accentuates the designers' concern to integrate the architecture with the functional requirements.

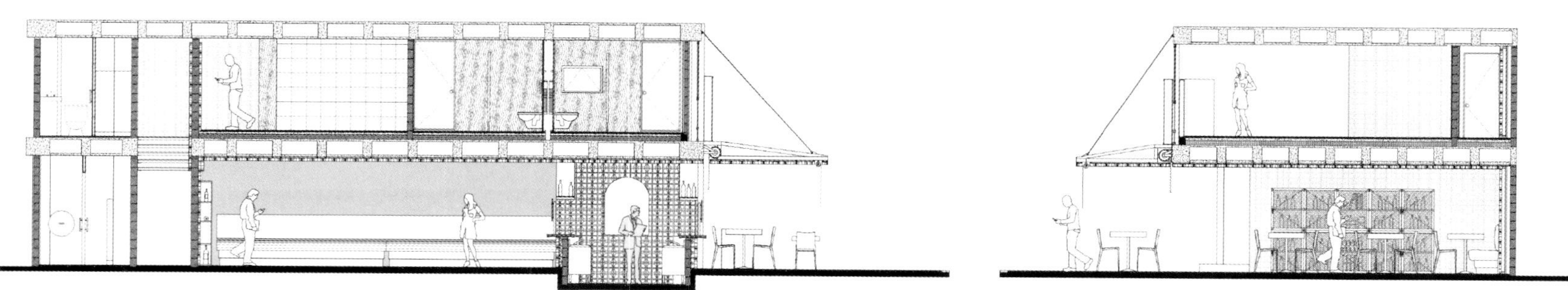

La Nonna是一家坐落于La Condesa的意大利餐厅。该项目是和DMG设计公司的设计师通力合作完成的。它旨在打破原有规划的束缚，从而使餐厅可以有效地利用200平方米的场地。餐厅的环境简洁、质朴。在施工过程中，设计师使用了当地的材料，地板由暗石制成，并铺设成人行横道的模样。墙壁和天花板上铺设着玻璃镜面，镜面上装饰着经过特殊处理并切割成块的红砖，墙壁上的镜面使空间得以扩大，从而制造出光线和阴影交织的效果。

餐厅内的设施既优雅又简单。餐厅中央是酒吧，木质的吧台后面有比萨饼炉，客人可以在吧台那里品尝美酒。经过一个优雅的发光楼梯就可以步入二楼，这一层是餐厅和厨房的所在处。在进入餐厅前，客人会看到一个夹层，里面摆放秋千、一些书以及一盏落地灯。这一切无不吸引着客人的眼球。

楼梯上的天花板以及主廊都被刷成黑色，目的是要与餐馆内部的红砖相呼应。照明灯隐藏于红砖之中，营造出动感十足的光影氛围。该餐厅的设计无不体现出设计师对于建筑物整体与功能理念的高度关注。

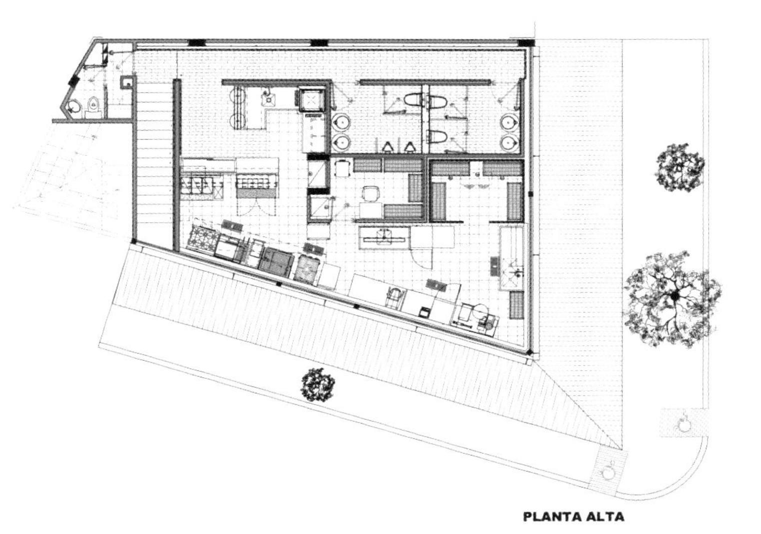

PLANTA ALTA

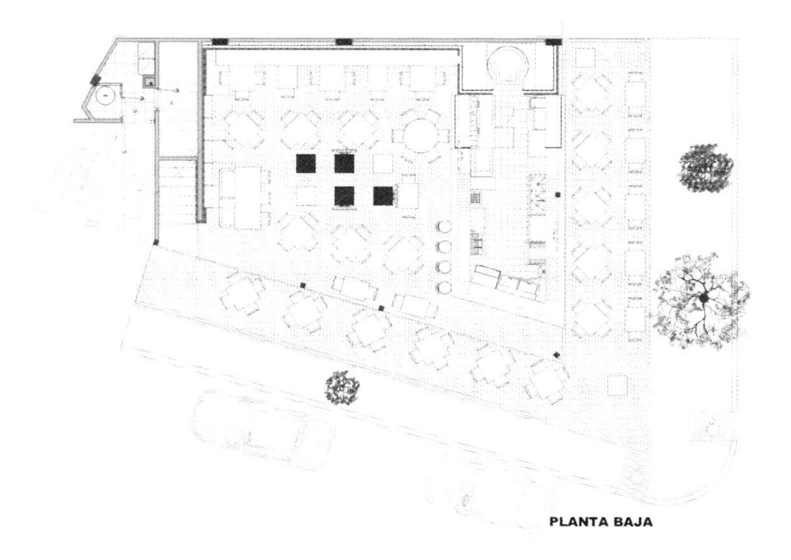

PLANTA BAJA

KARLS KITCHEN

Designers: Rob Wagemans, Melanie Knüwer, Charlotte Key, Erik van Dillen, Sofie Ruytenberg
Design Company: Concrete Architectural Associates
Client: E. Breuninger GmbH & Co.
Location: Marktstraße 1-3, 70173 Stuttgart, Germany
Area: 1,000 m^2 (including 40 m^2 terrace)
Materials: Oak Floors, Terrazzo Tiles, Floor Tiles Mosa, Acoustic Plasterboard Ceiling, Wallpaper
Photographer: Ewout Huibers

Karls kitchen is the new restaurant of the luxurious department store Breuninger in Stuttgart.
The kitchen is the free-flow restaurant's central element.
The entrance to the restaurant is characterized by large glass fridges with fresh products that also function as a display window.
To the left, guests can see exclusive sandwiches, fresh salads and homemade desserts being prepared. On the right, guests can take a peek into the warm kitchens, where traditional local classics, Asian and European specialities are prepared. The dishes, which change daily, are presented on a row of screens above the two kitchen counters.
Opposite the kitchens is the bar, a free-standing element between the free-flow restaurant and the seating area. The bar's function is twofold: on the kitchen side, there are cakes and pastries made in-store, while on the other side, fresh juices and beverages are offered.
The bar adjacent to the lounge area tempts guests to sit at the bar and enjoy a nice glass of wine, a glass of local beer or one of coffee specialities.
The restaurant is divided into three different seating areas, each with its own identity and character. First of all, there is the traditional seating area, which refers to Breuninger's history and the traditions within the department store, as well as in Stuttgart. The modern seating area represents the Stuttgart of today, and the lounge is the refuge and at the same time, the connection between the modern and the traditional seating areas.

Private Küche
Terrasse

Karl's餐厅是斯图加特市的奢华百货商店Breuninger内的一个新建餐厅。

餐厅是自助餐厅的中央元素。

餐厅的入口是一个大的玻璃冰箱，里面储藏着新鲜的食品，同时它也是一个展示橱窗。

向左转，可以看到正在制作中的独特三明治、新鲜沙拉和自制的甜点。向右转，可以看到温暖的厨房，里面正在烹饪着传统的当地菜肴，以及欧式和亚式的美食。菜品每天都会更新，在两个厨房吧台上方的一排屏幕上滚动展示每日新品。

厨房的对面是吧台，位于自助区和点餐区之间，是一个独立的元素。吧台有两个作用：告诉客人蛋糕和点心是店内自制的，另外还可以提供新鲜果汁和饮料。

吧台靠近会客区，似乎是吸引客人来吧台坐坐，享用一杯优质红酒、一杯当地啤酒或者餐厅自制的咖啡。

餐厅分成了三个不同的就餐区，每一个都有独特之处。首先有一个传统就餐区，主要是参考了Breuninger的历史、百货商店的传统以及斯图加特市的历史；第二个是现代化的就餐区，代表当代的斯图加特市；第三个是会客区，这是一个休息的场所，同时也是现代和传统就餐区的过渡区。

concrete

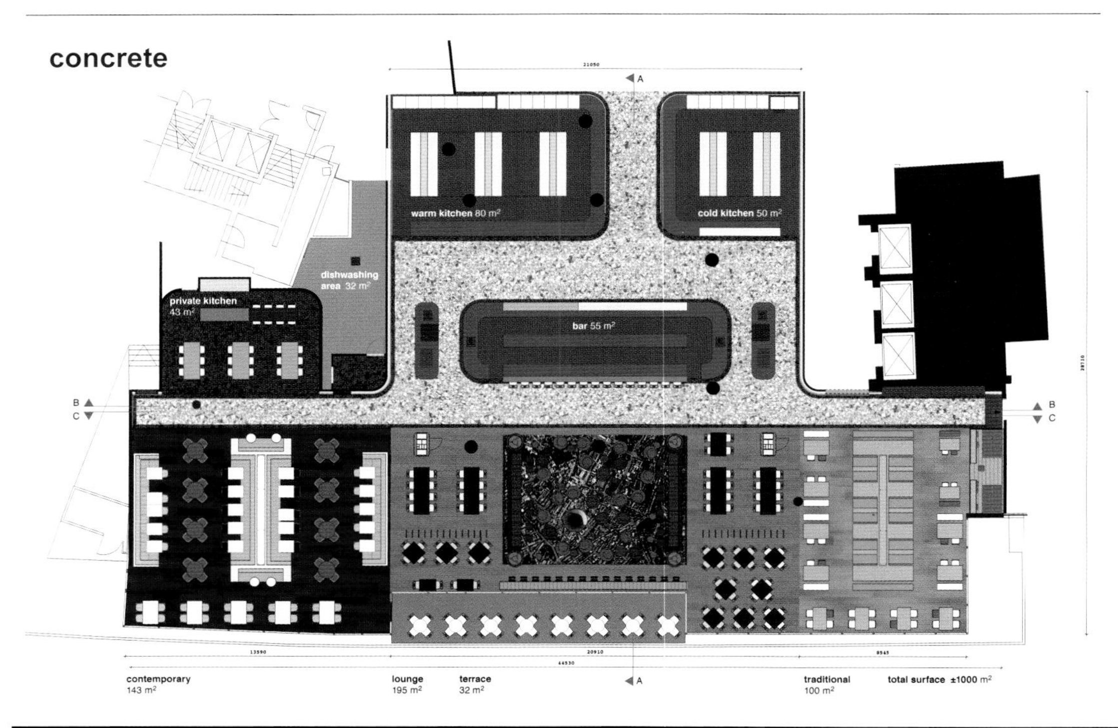
warm kitchen 80 m²
cold kitchen 50 m²
private kitchen 43 m²
dishwashing area 32 m²
bar 55 m²
contemporary 143 m²
lounge 195 m²
terrace 32 m²
traditional 100 m²
total surface ±1000 m²

Ausgang ▸
Toiletten ▸

Ausgang ▸
Toiletten ▸

YILAN XINYUEGUANGCHANG YUYAN

Designer: Frankie Yu
Design Company: Gure-Design

The designer takes the natural elements into the interior by using creative ingenuity, so that it can give more tension to the whole space. Using black and brown wall mirror as the main focus of the material can reflect the interaction between the chef and customers vaguely. In this way, the designer adds the entire space a dinning atmosphere which is more penetrating and lively. Because of the feature of glass partition and curtain in soft compartment, the space can be used in a flexible way. Floor level changes and indirect light both lead to a visual extension.
Clean lines combined with a bold use of color make each point has a different visual surprise. Lighting is also an essential element to make a difference. It's really a feast enjoying food in such a comfortable environment.

御宴
和風鐵板燒料理

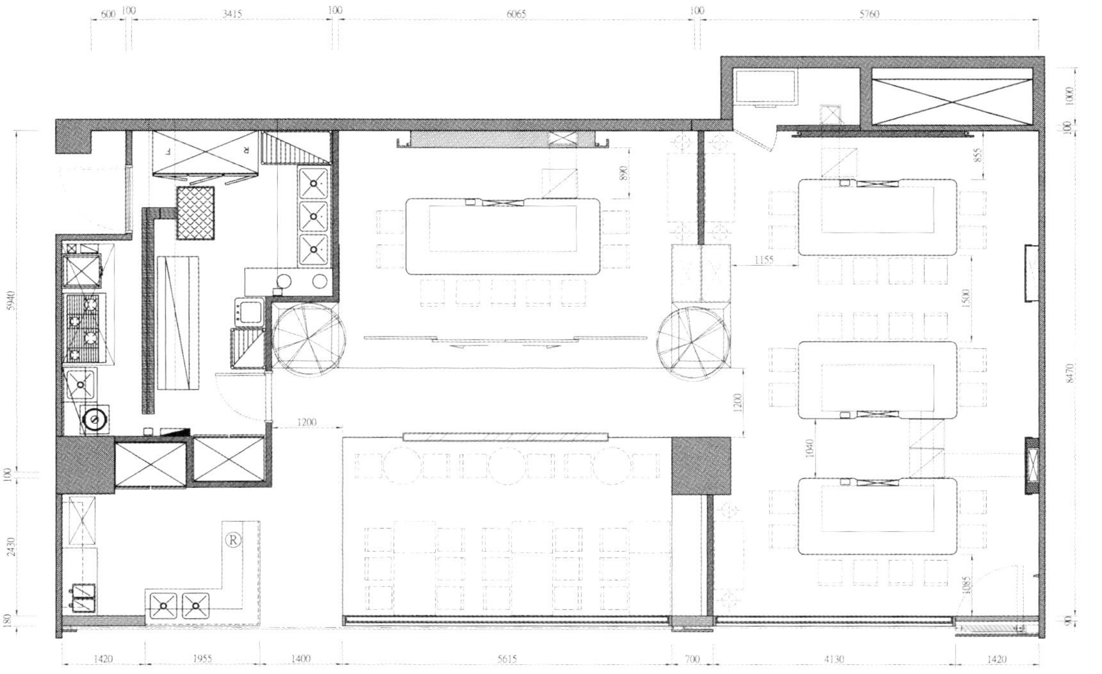

设计师以匠心独具的巧思,将自然的元素带入了室内，令整个空间更具有张力。以黑色与茶色镜面作为重点墙面的主要材质，隐约反射出了客人与厨师间的互动，更增添了整个空间的穿透性与活泼的用餐气息。

以玻璃隔屏与门帘的软隔间特性，使空间运用更加灵活，地坪的高低变化、由间接光源造成视觉的延伸，以及利落的线条和大胆的用色，使每个角度都有不同的视觉惊喜。

照明也是营造气氛不可或缺的一大功臣，在舒适的环境下享用美食，真是美食与设计的飨宴。

KAREN TEPPANYAKI, SOGO

Designer: Frankie Yu
Design Company: Gure-Design
Area: 40 m^2

The designers use the color black, red and natural wood to show the contemporary and gorgeous charm of the space. By doing this, the whole space becomes prominent and distinctive at once. Different materials mix with each other lay bare to an inquiring eye in the space.

The space uses the ㄇ-shaped iron mesa as the primary focus, while the composition of the wall stands as the big wooden tree folded into three-dimensional surface. The rear main wall has a totem tea mirror formed by black Sands paint which amplifies the sense of space successfully. The wooden ceiling is designed into a fillister shape, which increases the dynamic elements of the space as well as showing the beauty mixed with .rough primitive elements and modern conflict.

Brown leather high chair, slight metallic luster and metallic brick show modern and modest luxury. Designers embedded the pot within the table carefully, which is users-friendly. A cashier machine displays in the corner of the counter. The "Moving Line" of the space is comfortable, with the top of the counter displaying the price list which appears in the form of Gold Braille on the mirror. That design integrates with the style of the space perfectly and also functions as accommodating amazingly.

"Enjoy design" and "tasting food," are the beliefs the space try to convey.

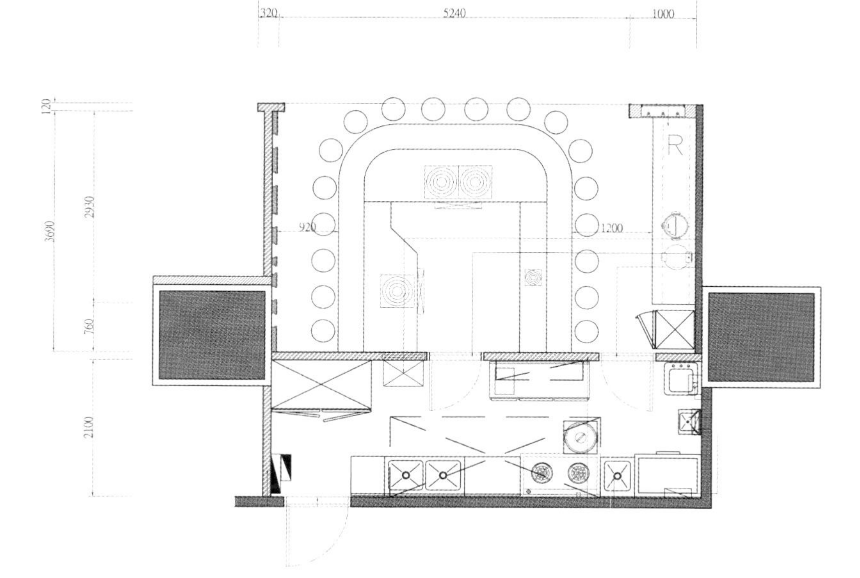

设计师运用黑、红和自然木纹色调，展现出现代、华丽的迷人风采，整体空间立刻变得突出、与众不同、异材质的、混搭技巧在空间中展露无遗。

该空间以П字形的铁板铁台面为首要重点，后方主墙上镶嵌着印有图腾的茶镜，成功地增强了空间感，木纹美耐板则利用格栅凹槽设计的天花板增加动感元素，呈现粗犷原始与现代利落冲突的美感。

咖啡色人造皮的高脚椅、微微的金属光泽和地坪铺陈的金属砖，流露出时尚、现代的低调奢华，设计师细心地将锅具内嵌入台面，使用上更为顺手、方便，柜台内角陈设收款机，店内动线安排流畅，上柜为价目表展示区，金卡点字在茶镜上的呈现，与店内风格完美融合，还兼具了实用的收纳功能。

"享受设计"和"品尝美食"，是该空间所传达的理念。

AUTOSTRADA

Designer: II BY IV Design Associates
Location: Autostrada (Italian for "Highway"), at 3255 Rutherford Road in Vaughan, Italy
Area: 585 m^2

At the entrance in Autostrada, the leather-finished statuario marble desk thrusts in contrast toward guests, from a gleaming, high-luminosity beveled glass wall panel. To the left, the feature wall of banquette settees looms skyward from seated patrons, and is affixed with an intricate Pirelli tire tread. Lit from the top and bottom, the tread wall is luminescent, bold and graphic. Striking black and white city-grid fabric upholsters the banquette settees.

Balancing the highway theme to other elements was home kitchen aspects of white oak butcher-block tables and contoured Eames chairs with dowel wood bases. In the Pizza Bar, behind the statuario marble work area the pizza oven is clad warmly from ceiling to floor with a blaze of mini-mosaic style tiling in brilliant red-orange that gives the impression that the entire wall is of fire.

The square footage of the private dining room is modest; however, the ceiling is towered well above proportion. To achieve balance, the designers dropped the ceiling, incorporated a dropped installation of baton lighting and introduced rear-view mirrors to ground guests' gaze back down into the room.

The center of restaurant — the Sunken Dining Area is encircled by a custom lacquered screen of inverted trapezoids. The centerpiece is surrounded by a three-quarter moon of butcher-block tables. Here, the floor treatment departs from Terazzo tiling to white oak.

In the main bar,the quality wine cellar is elevated to be in full view with a backlit chrome and white oak encasement. Beneath this is another textured bar top, punctuated by a row of lighting pendants, and supported by a brilliant orange-red beveled glass base that comprises the color of the room.

餐厅入口处，摆放着一个意大利白色大理石桌，与闪闪发光的高坡度玻璃幕墙形成强烈对比。左侧是宴会厅的功能墙，墙面紧靠客人用餐席，墙上贴有错综复杂的倍耐力轮胎胎面，从顶部到底部，此墙面给人的感觉是发光、大胆、棱角分明。宴会厅内部摆设了一个引人注目的、黑白相间的、外形酷似城市电网结构的长沙发。

充分体现高速公路主题的就是内置白橡木桌子以及木质埃姆斯椅子的家庭厨房。厨房内有一个比萨吧，在吧台后面的工作区，表面包钢的比萨烤箱从天花板一直延伸至地面。

厨房四周的墙壁上铺设了迷你马赛克样式的橘红色瓷砖，给人一种“火墙”的感觉。私人餐厅的空间大小非常恰当，然而天花板的比例却偏高，为了保持平衡，设计师放弃了对于天花板的设计，而是在上面挂了吊灯，并铺设了玻璃墙面，使餐厅极具空间感。

餐厅的中心是Sunken就餐区，被倒梯形的屏幕所环绕，内部是白橡木地板以及月形的桌子。

主酒吧里装有优质红酒的酒窖区被高高架起，能够看到黄色的背光灯和白色橡树装饰的设施。下面还有一个纹理不同的吧台，与上面悬挂的一排吊灯遥相呼应，这与闪亮的橙红色斜面玻璃融为一体，使整个房间的颜色协调一致。

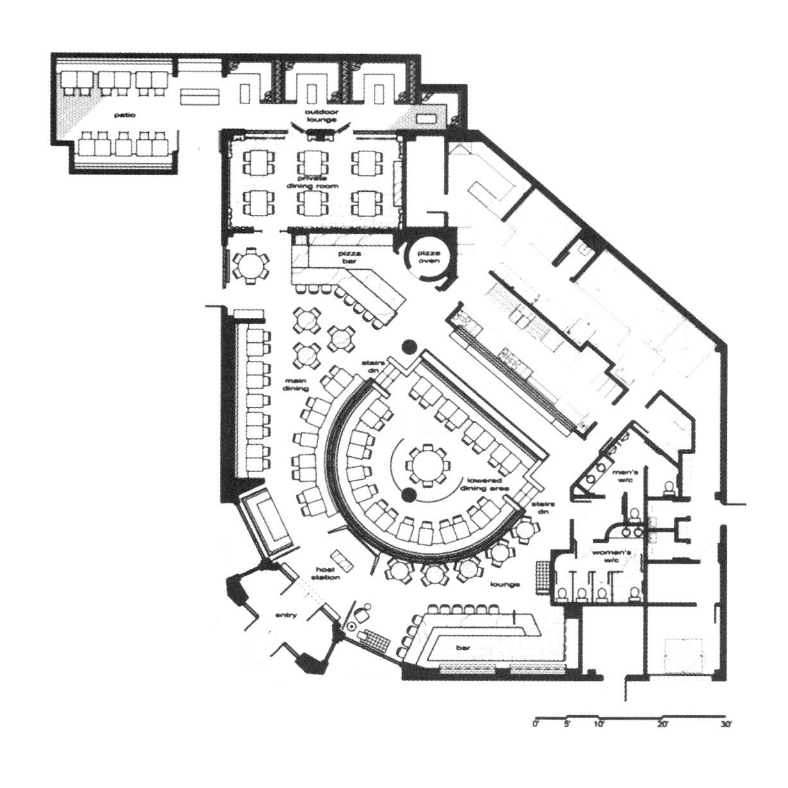

CHAIRMAN'S SUITES PRIVATE MEMBERS LOUNGE

Designer: II BY IV Design Associates
Location: Air Canada Center, Canada
Area: 418 m^2

The designers chose classic, simple, and luxe materials
Entering the lounge, one moves through "the Portal": a short tunnel, lined with vintage wines of the month, inset in bronze mirror-clad walls and behind glass. Graphically stunning ceiling detail with concealed lighting gives the Portal a feeling of airiness. Apart from its aesthetic function, it divides the lounge into two sides.
On each side, banquette seating with rosewood tables accompanied by cognac-colored leather and polished bronze chairs line the outside of the space and curve around to the centre mark. The outer wall is clad with polished bronze rods in a weave, laid against an espresso backdrop. Underfoot, vanilla marble tile punctuates the room with quality, and is also understated.
Finishing off the outer walls rests private booths. Clad in a black-painted, beveled glass surround in a matte and gloss texture, the inner wall is espresso leather-paneled and is lit by a polished bronze "fin" lighting fixture.
In the center on each side, ample booth seating inside massive and regal rosewood surrounds, sit under four-armed, spider like lighting features, dropping from the ceiling the polished bronze fin lighting. This section is anchored by dark hardwood.
The most striking feature in the main bar is the magnificent wall cladding. An asymmetrical, multi-faceted grouping of bronze mirror, etched bronze mirror and black painted black glass provide extraordinary visual interest and an occasional sparkle of reflection from lighting or other parts of the room.
Private booths are located at the outer reaches. These booths have custom vintage inspired fin lighting above, and are distinguished by floor-to-ceiling, curved rosewood surrounds with uniquely designed, bronze, overlaid screens laid against the inside of the surrounds.

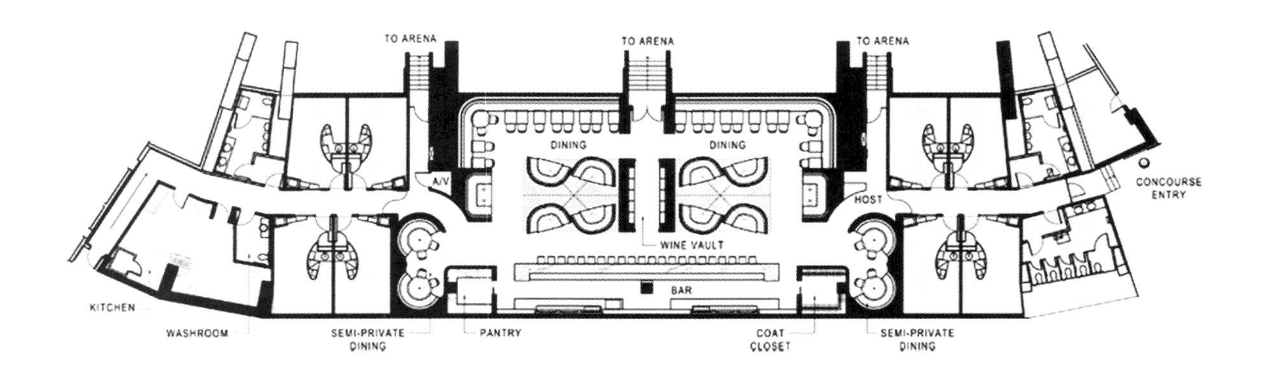

设计师选用了经典、简洁和奢华的材料。

进入会客厅，就能看到入口：一个短的通道，两边摆放着本月的精品葡萄酒，墙壁上镶嵌着铜色的镜子。漂亮的图形化天花板和隐匿的照明给人一种微妙的感觉。这个入口除了有美学的功用之外，还将这个会客厅分成了两部分。

每一个部分都配备了紫檀木的桌子和配套的餐椅。餐椅的材料选自法国干邑白兰地颜色的皮革和抛光铜。餐椅沿大厅外侧摆放，弯曲排列，直到大厅中心。外墙覆盖着发亮的铜色细杆，呈交织状的排列，后面映衬着Espresso咖啡色的背景。在脚下，香草的大理石饰面砖更加突出了房间的质感、低调。

外墙之外，就是私人包间。黑色喷漆的斜面玻璃贴饰在包间外，亚光质感和亮光的光彩交相辉映，内墙是Espresso咖啡色的皮质，铜色的fin照明装置进行照明。

每一部分的中央是宽敞的包间，围拢在巨大的、富丽堂皇的紫檀木之间，四臂的支架照明灯采用铜色的fin 照明装置，从天花板垂坠下来，照亮整个包间区域。深色硬木是这个区域的主要建材。

主吧台最引人注目的是华丽的墙体装饰。铜色镜子与蚀刻铜色镜子的不对称、多立面的搭配，以及黑色喷漆玻璃，不经意间反射出灯光和室内其他物品，给人以极强的视觉震撼。

私人包间位于会客厅的外侧区域。这些包间配备了定制的fin照明灯，设计独特的屏幕悬挂于会客厅的中央，通体而上的弧形紫檀木凸显出包间的高雅。

THE
CHAIRMAN'S
SUITE

CIENNA RESTAURANT

Designer: Antonio Di Oronzo
Design Company: Bluarch Architecture
Location: New York, USA
Area: 307 m^2

Cienna Restaurant is a venue in the heart of Astoria Designed for Chef Eric Hara, this venue offers fine dining through a contemporary American menu with influences from Mediterranean, French and Asian cuisines.

This project is a conceptual exercise on lightness as a philosophical category.Greek philosopher Heraclitus defined lightness as a fundamental mode of existence with absence of existential burden. This venue formalizes the flexible, the weightless, the mobile, the connective vectors as distinct from structure offering experiential lightness.

The layered lightness of the design exemplifies the fleeting, the ephemeraland the whimsical excitement of the present moment. A sinuous pattern of mahogany wood members is punctuated by plush upholstered panels in a saturated red. The sheer rose curtains behind this latticework and the mirrored ceiling multiply a weightless, mobile sense of space, as they vibrate in reflections and translucency. The bar is a lighted prism of pink onyx which seemingly floats diaphanous while diffusing a soft glow.The chairs designed by Charles and Ray Eames and Eero Saarinen organically embrace the diners, as their experience is sensually lifted by the luscious food and the warm environment. The lighting design is integral to the seamless performance of all design systems. A state-of-the-art LED system of fixtures allows for total flexibility of ambience and mood.

Cienna餐厅位于纽约阿斯托利亚市区中心。 Bluarch Architecture + Interiors设计工作室承接了该设计。该餐厅供应时尚的美式菜品，这些菜品深受地中海式、法国和亚洲菜式的影响。

该设计力求表现由希腊哲学家赫拉克利特定义的“轻盈”的设计理念，即，餐厅空间灵活多变，轻盈灵动，有别于传统的架构式装修，能够让客人实实在在地体验“轻盈”的氛围。

该设计的纤薄的层状装饰让人感受到了时光易逝。镶嵌着桃花心木纹饰的、弯曲的大红软板，格子框架后面透明的玫红色窗帘和镜像天花，以及多处映射的效果和半透明的装饰更加强化了空间这种“轻盈”的概念。吧台是浅粉的缟玛瑙表面，在漫射的灯光下似有水波在流动。坐在Charles and Ray Eames 和 Eero Saarinen设计的舒适的座椅里，客人在优美的环境里享用香甜的食物，似乎有一种“食在人间不慕仙”的感觉。这种轻灵的设计对于无缝连接的装修是不可或人缺的。现代技术的LED 灯为餐厅“轻盈”的整体设计增色不少。

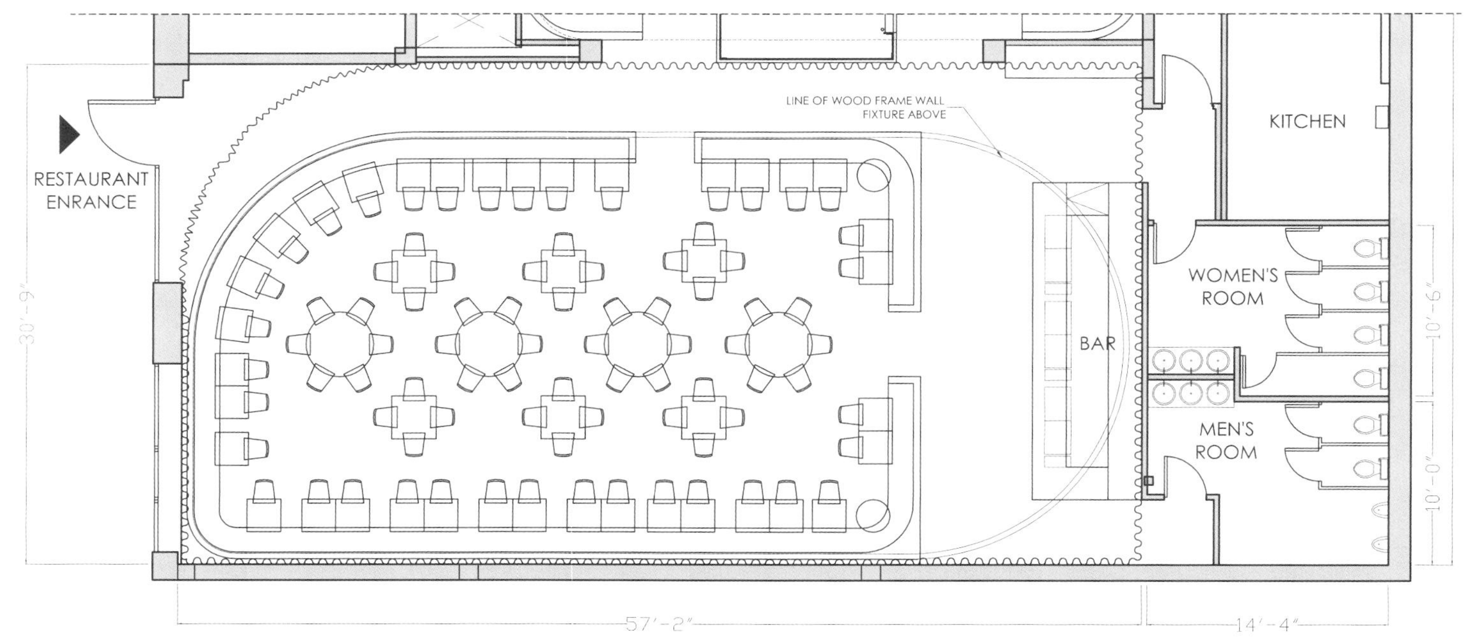

GMOA KELLER

Designers: Michael Anhammer, Christian Ambos, Harald Höller
Design Company: SUE Architekten
Location: Vienna, Austria
Photographer: Hertha Hurnaus

The recently renovated Viennese restaurant, GMOA Keller has been an iconic part of the cityscape since 1858, and is a time-honored institution known for its authentic charm. Because of this, the business has been always running successfully, and the audiences and artists of nearby Konzerthaus and Akademietheater guarantee full tables, therefore, the owner decided to expand the restaurant under-ground. This meant that the team at local architecture firm SUE Architekten was challenged with both adding a sub-level and keeping the space's authenticity and charm intact. The response was a contemporary, calm and spacious dining area, allowing guests to enjoy excellent Viennese cuisine while offering space for regulars to have their beer at the classic Resopal table. The design is uncomplicated, the brick arches were simply whitened, the wooden floor and panelling were stained dark, and the theatre curtain rails allow the rooms to be divided into zones and niches with curtains by Kenzo. Light fixtures by Moooi are mobile and can hang over tables creating privacy, or if the light isn't need, they can easily be pushed aside and the room is again at its full dimension.

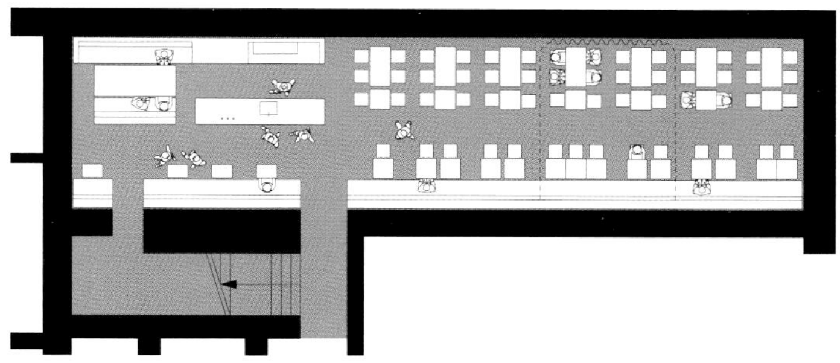

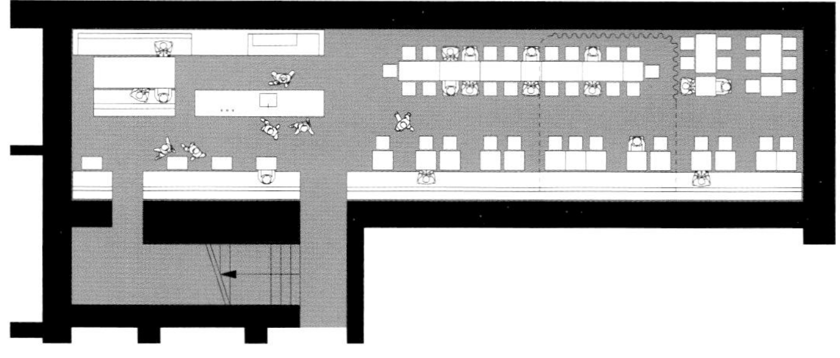

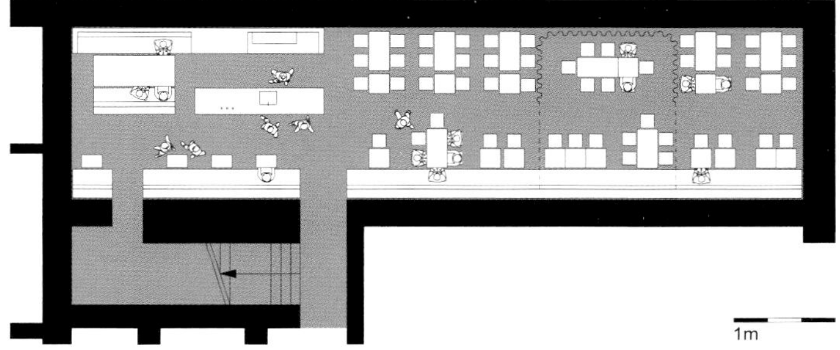

***SUE** ARCHITEKTEN*

GMOA Keller

最近装修的维也纳餐厅GMOA Keller自1858年以来就作为城市景观的一个标志性建筑伫立于此，可谓历史悠久。其魅力已人所尽知。因此，生意很红火。附近的住户和艺术家经常光顾，所以常常客满。餐厅经营者希望在地下室扩建餐厅，这就意味着设计公司SUE Architekten ZT KG，将迎接前所未有的挑战——既要增加空间层次，又要保持空间的真实性和魅力性。于是，设计了一个典雅、宽敞的用餐区，这里可供客人享受地道的维也纳美食或品尝经典的Resopal啤酒。幕布的使用使房间可以被划分成众多的小区域。

此餐厅的装饰设计并不复杂。砖拱是简单的白色，地板是木质的，镶板是暗黑色的。照明灯是荷兰Moooi创意灯饰，悬挂在餐桌上方，并且可以移动，如果不需要照明，这些灯具很容易就可以被移动到旁边，餐厅就又重现了完整的空间。

HARMONIE H AT STATE HIMEJI BANK

Designers: Hideki Kureha, Noriaki Takeda, Ikuma Yoshizawa
Design Company: Tao Thong Villa Co., Ltd. + Process5 Design
Area: 1,976.88 m^2
Location: 47 Minamimachi Himeji-city Hyogo, Japan
Photographer: Tao Thong Villa Co., Ltd.

The designer received a request to undertake a project to renovate a building which has coexisted as a bank amid the castle town of Himeji for half a century, located near Himeji Station, into a wedding hall. According to the plan, the designer decided to create two banquet halls, onechapel and an attached space. The chapel named "CHAPEL DE CORDULA" was secured through structural reinforcement of a large nine-meter-high space from the first basement floor to the third floor slab.

Giving the impression of stained glass, a total of 210 custom-made pendant lights consisting of 7 colors were suspended. The space encompassed by beautiful lighting is a stadium shape so that the guests may look on the bride and groom and their moment of exchanging vows.

The banquet halls are on the first and third floors. The walls of the first floor banquet hall "HANNELORE" are decorated with various paintings, primarily oil paintings. The third floor banquet hall "ARTDUCUT" has as its theme elements related to food such as cutlery and flame, and has a lively atmosphere with the operation of a restaurant as well. The banquet hall and the kitchen have a visually open plan with only a sheet of glass partitioning them. From the guest seats, the entire kitchen can be overlooked, being able to enjoy food while sensing the performance of the chefs.

Café "LACER DE YUIT " located on the first floor was planned so that the customers can enjoy the free atmosphere, open to the town.

The entire facility is decorated with paintings and sculptures by domestic and foreign artists, a pleasant welcome to the guests.

The letter "H" contained in the name of facility "HARMONIE H AT STATE HIMEJI BANK" brings to mind many feelings such as "Happy," "Human,""Harmonie," and "Himeji," all starting with the letter "H."

This ceremony hall which emerged in the historical castle town of Himeji, has become a facility which bonds people, makes visitors happy and aims to achieve harmony with the people of the town.

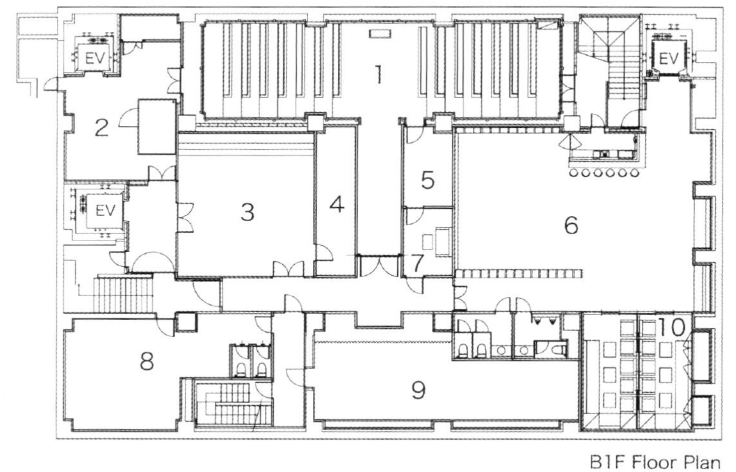

B1F Floor Plan

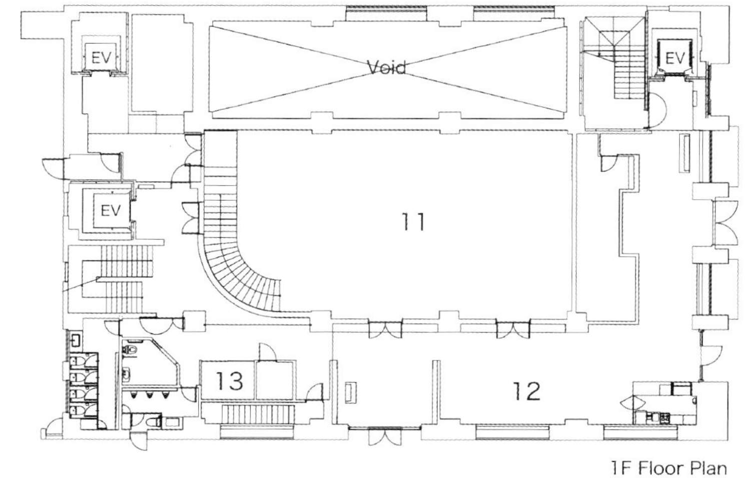

1F Floor Plan

1.Wedding chapel "CHAPEL DE CORDULA"
2.Flower staff room
3.Photo studio "TAO THONG VILLA HIMEJIDO"
4.Video device room
5.Staff space
6.Foyer "LIBLARYRALBIL"
7.Waiting space
8.Staff room
9.Bridal salon "BRIDAL SALON FELICIE"
10.Anteroom for relative

11.Banquet "HANNELORE"
12.Café "Lacer de yuit"
13.Smoking room

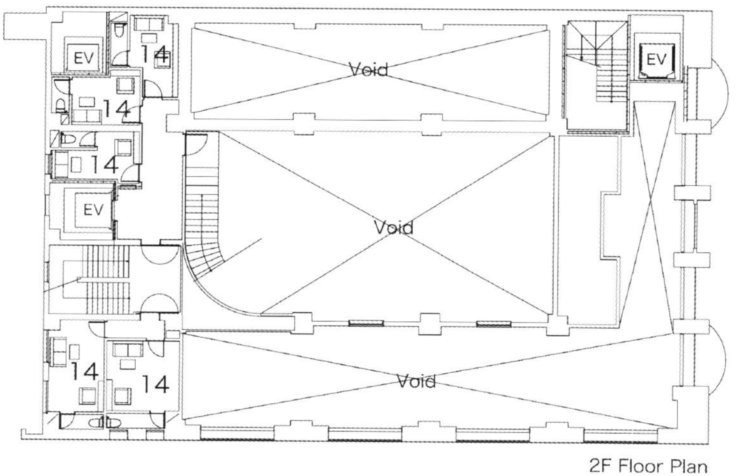

2F Floor Plan

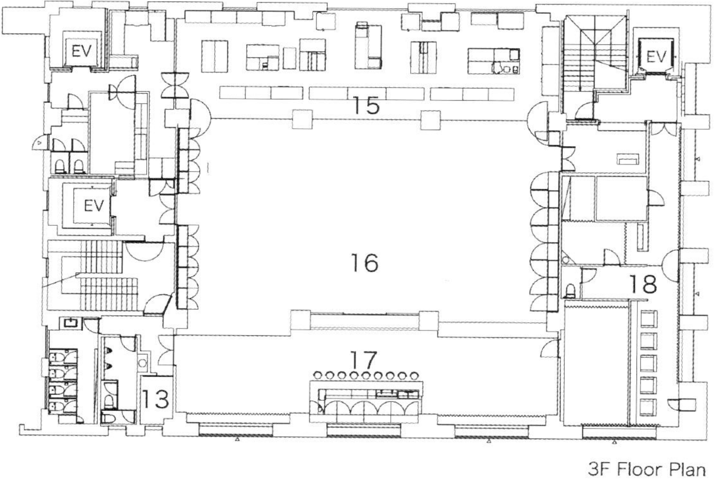

3F Floor Plan

14.Brides room
15.Kitchen "TASTINGS"
16.Banquet "ARTDUCUT"
17.Bar "HIMEJI CLUB"
18.Beauty salon

设计师接到一个改造旧楼的工程项目。该建筑物先前是一家银行，在Himeji城已有半个世纪的历史，位于“姬路”附近，通向结婚礼堂。根据规划，设计师决定建造两个宴会厅、一个教堂和一个附属空间。教堂的名字为“CHAPEL DE CORDULA”，设计师希望通过加固框架结构，使这个9米高的建筑能够稳固地伫立于此。

教堂内的彩色玻璃令人印象深刻。除了悬挂在天花板上的210个特别订制的七色吊灯，整个空间的墙壁上也挂满了美丽的灯饰。整个教堂被设计成一个体育馆的形状，这样在举行婚礼的时候，客人很容易就能看到新娘和新郎，以及他们在上帝面前许下誓言的全过程。

宴会厅在一层和三层，一楼的宴会厅“HANNELORE”墙壁上装饰着各种绘画，主要是油画作品。三楼宴会厅“ARTDUCUT”主题内容主要与食物相关。所以，墙上画了许多餐具和火焰，以及部分餐厅食物的制作流程。宴会厅和厨房仅有一个玻璃墙间隔，从而营造出通透的氛围。客人坐在椅子上品尝美食的同时，也可以看到厨师在厨房里烹制菜肴的全过程。

“LACER DE YUIT”咖啡厅在一层，这样设计旨在令客人能够随意享受室外空间。

整个空间都摆满了国内和国外的艺术家设计的雕塑和装饰画，给人一种愉快的感觉。字母“H”包含在“HARMONIE H AT STATE HIMEJI BANK”里，给人一种感觉：“快乐”、“人权”、“和谐”和“姬路”，因为以上所有单词都是以字母“H”开头的。

这个建筑伫立于历史悠久的姬路城堡镇，俨然已成为当地的标志性建筑。

SEAFOOD BOAT OF TANG PALACE

Designer: Chang Yung Ho
Project Architect: Lin Yihsuan
Design Team: Yu Yue, Wu Xia, Suiming Wang
Location: Hangzhou, China

The restaurant selects composite bamboo boards as the main material, conveying the design concept of combining tradition and modernity. In the hall, to take advantage of the story height, some booths are suspended from the roof, creating an interactive atmosphere between the upper and lower levels, thus enriching the visual enjoyments. The original building condition has a core column and several semi-oval blocks which essentially disorganized the space. Hence, the design intends to reshape the space with a large hollowed-out ceiling which is made from interweaved thin bamboo boards and extends from the wall to the ceiling. The waved ceiling creates a dramatic visual expression within the hall. The hollowed-out bamboo net maintains the original story height and thereby creates an subtle interactive relation between the levels. The designers also wrapped the core column with light-transmitting bamboo boards to form a light-box, which transforms the previously heavy concrete block into a light and lively focus object.

The entrance hall also follows the theme of bamboo. The wall is covered with bamboo material which follows the original outline of the wall, turning it into a wavy curve. In this way, the curve not only echoes the hall ceiling but also performs a guiding function for customers, making guests feel fresh at the entrance.

The booths on the first level are relatively bigger and share the features of expanded bamboo net from the wall to ceiling and ornamentally engraved wall surfaces. Meanwhile, the different folding angles and engraved patterns make each booth different from one another. The booths above the interlayer on the south are smaller and feature a special waved ceiling pattern and simple bamboo wall surface, which creates interesting and spacious room features. The key design concept of the space is that the suspended booths are connected with suspended bridges and sideway aisles. The semi-transparent wall provides a subtle relationship between the inner and outer spaces, offering guestets with a special spatial experience.

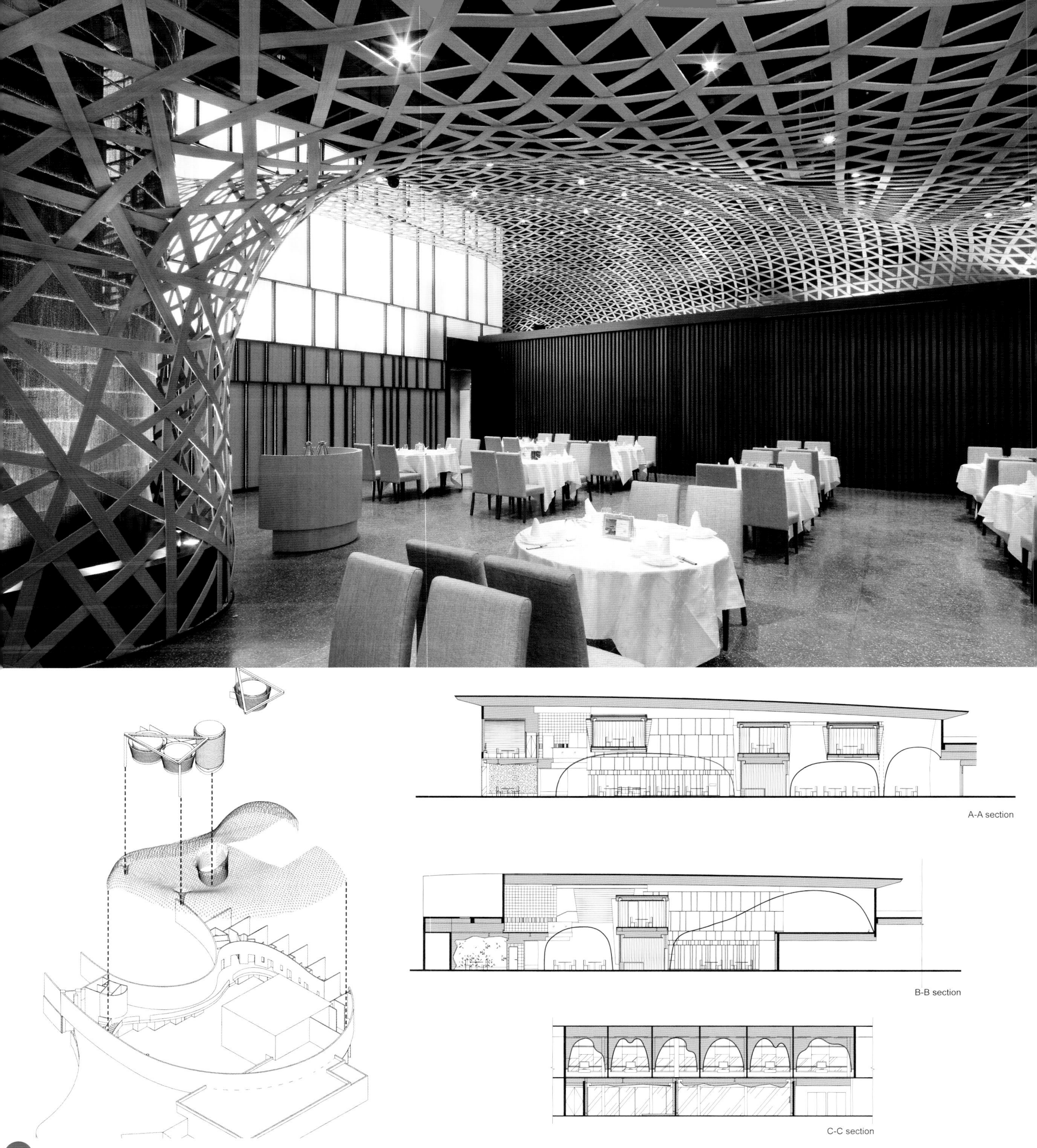
A-A section
B-B section
C-C section

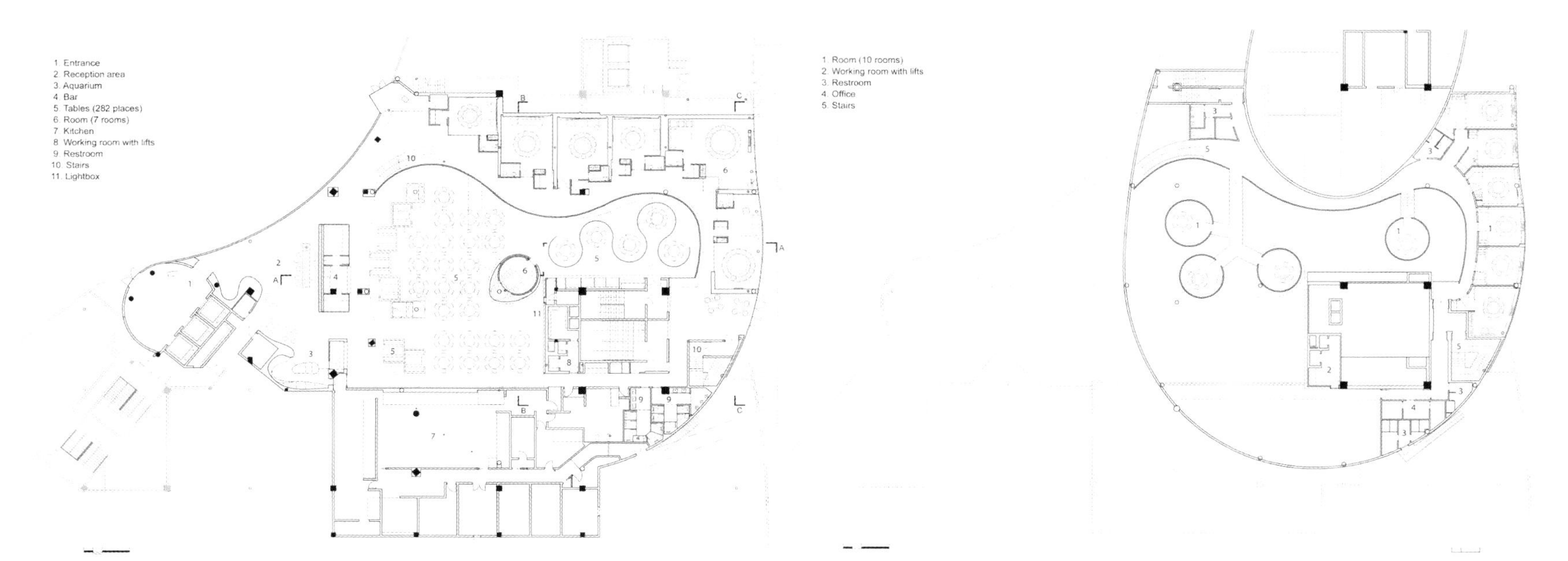
1. Entrance
2. Reception area
3. Aquarium
4. Bar
5. Tables (282 places)
6. Room (7 rooms)
7. Kitchen
8. Working room with lifts
9. Restroom
10. Stairs
11. Lightbox
1. Room (10 rooms)
2. Working room with lifts
3. Restroom
4. Office
5. Stairs

本案选用复合的竹板为主材料，强调传统与现代相结合的设计主题。利用大厅中原有的层高优势，设计师将部分包间悬吊于顶上，创造出高低层次的趣味性，丰富了空间的视觉感受。大厅中心巨大的核心筒和侧边悬挑的半椭圆形体块使空间显得零碎杂乱，设计师采用一片由薄竹板编织、从墙面延伸至天花板的巨大透空顶棚，将空间重新塑造，构筑了大厅里戏剧般的场景。而视线穿过透空的竹网，不仅保持了原有的层高优势，亦使得上下层有了微妙的互动关系。在原有的核心筒外，设计师以透光竹板包覆四壁形成灯箱，使原本沉重的混凝土体量变为空间中轻盈的焦点。

入口门厅延续了竹的主题。墙面覆以竹材，并顺应原有的墙体处理成波浪状流动的弧面，除了与大厅的顶棚相呼应外，也具有空间导引的功能，使客人一进入餐厅就感觉耳目一新。

一层的包间较大，从天花板到墙面的折板和两侧镂花透光墙面是共同的基础语汇，但每间各自不同的折板角度和镂花图案，令各个房间有了彼此相异的面貌。南侧夹层上方的包间略小，借由特殊的曲面顶棚造型和简洁、单纯的竹材墙面，营造空间的趣味感并凸显其大气之风。至于作为空间重点的悬挂包间，以空桥和侧边走道连接，半透明的墙面形成隐约的内外关系，使人不论在其内或外，都能够享受特殊的空间体验。

MOMBÓ

Designers: NMD|nomadas, Farid Chacón, Francisco Mustieles, Claudia Urdaneta
Design Company: NMD l nomadas
Construction Supervisor: Ricardo Prieto, NMD l nomadas
Location: Maracaibo, Venezuela
Area: 513.03 m²
Photographer: Luis Ontiveros

Mombó located in Venezuela, perhaps the clearest and simplest proposal on the creation of a restaurant and bar interior design is the adoption of the expiration of the environment "show" that has deprived over the past decade,. Therefore, it is important to call for reflection on the urgent quest for domestication to the possibilities of its application in creating new spaces for a user saturated, tired and exacting. The show is supplemented by the "home" or by the ideal of home transformed into public space, a space for intimacy, enjoyment with friends and relaxing break, Mombó therefore seeking to embark on a private dimension, welcoming and free of claims. A true interior design showing what freedom really means by NMD.

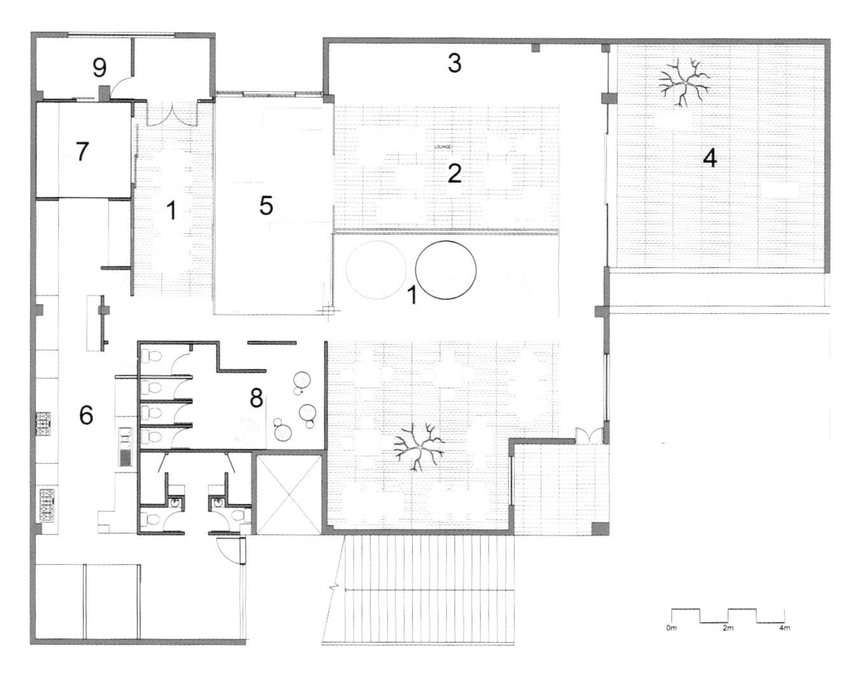
9
3
7
2
4
1
5
1
6
8
0m
2m
4m

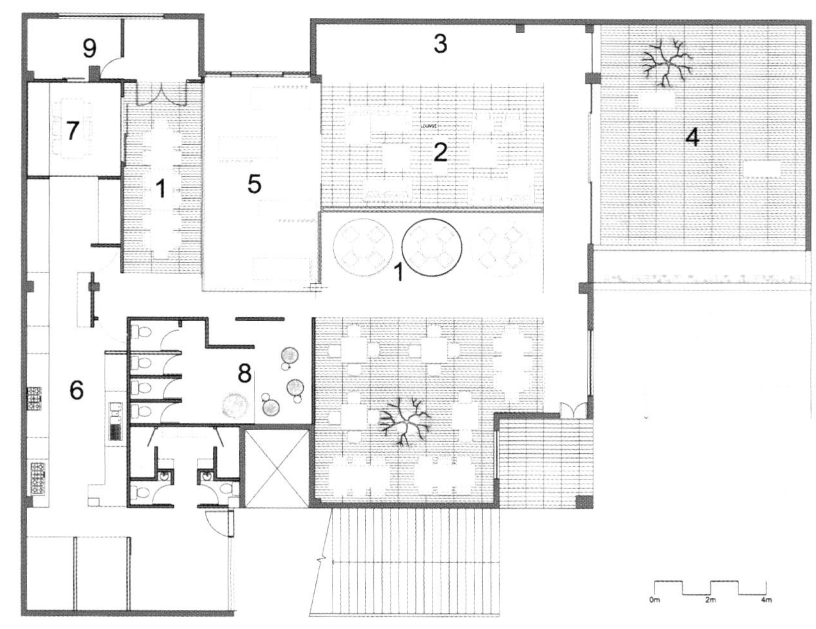
9
3
7
2
4
1
5
1
6
8
0m
2m
4m

Cafe

Mombó位于委内瑞拉，其简洁的内部装修源于对十年来的“环境展”的完美借鉴，所以，唤起意兴阑珊的人们对创造新环境的渴求尤为重要。“展览”有“家”和“公共空间”的元素，是一个既私密又可以交朋会友的空间。所以，Mombó的设计力求私密有度，轻松、自由。

BLYTHSWOOD SQUARE

Designer: Jim Hamilton
Design Company: Graven Images
Architect: Ron Galloway Architects
Location: Blythswood Square, Glasgow, UK
Photographer: Renzo Mazzolini

Graven Images were interior designers on Blythswood Square, Glasgow's only five-star hotel, the former headquarters of the historic Royal Scottish Automobile Club. The designers designed the one hundred bedrooms, world-class spa, screening room, the rally bar, the main bar and restaurant, meeting and event spaces, the salon and reception. They have also designed all of the furniture throughout the whole of the project — a brand new range of bespoke furniture pieces including dining chairs, lounge chairs, tables, all bedroom furniture, and all of the bathrooms. Besides, there are a range of bespoke lampshades and light fittings and bespoke carpets and rugs throughout.

The club was one of eight official starting points of the 1955 Monte Carlo Rally, which returned to its starting point in 2011 to celebrate its 100th Anniversary. The designers specified a wide range of Harris Tweed throughout including some patterns that haven't been used in sixty years. They spent time at the mill with the weavers, to decide on the range.

Blythswood Square has received much acclaim in the hotel and leisure industry, and was named on the Condé Nast Traveller Hotlist of places to stay in 2010. The hotel was also awarded Scottish Hotel of the Year and the Scottish Hotel Design award at the Scottish Hotel Awards 2010. The designers are thrilled that one of the recent hotel projects has been recognised by the industry in this way.

Graven Images是一家位于Blythswood广场的室内设计公司。Blythswood广场上有格拉斯哥唯一的五星级酒店，此酒店的前身是历史悠久的苏格兰皇家汽车俱乐部。设计师设计了100间客房、世界级水疗中心、放映室、集会吧、中心酒吧、餐厅、会议室、活动室、沙龙以及接待处。此外，还有符合整个项目风格的所有家具。这是一套全新的品牌定制家具，包括：餐椅、休闲椅、餐桌、所有的卧室家具和浴室设施、灯罩、灯具和地毯。

该酒店前身——苏格兰皇家汽车俱乐部是1955年8个官方蒙特卡洛拉力赛点之一。2011年是其成立100周年。为了庆祝这一时刻，设计师精心挑选了一些60多年都未曾在这种场合下使用的哈里斯花。

该酒店目前得到了酒店业和休闲业的普遍欢迎。2010年，它在康德纳斯旅行者推荐的热门名单上榜上有名，并且赢得了2010苏格兰酒店设计奖。该酒店的设计已经获得业界的广泛认可，这令设计师倍感骄傲。

BARBICAN FOOD HALL AND BARBICAN LOUNGE

Designer: Helen Hoghes
Design Company: SHH
Area: 450 m^2
Materials: Mild Steel, Ceramic, Glass, Brick
Photographers: Gareth Gardner, Caroline Collett

SHH served both as interior and branding designer on the project, which encompassed both the former 450 m² ground floor café, now a restaurant and shop, re-branded as "Barbican Foodhall", and the first floor bar and restaurant - now "Barbican Lounge", as well as the outdoor terrace spaces for both venues.

"Our overall approach", explained Helen Hughes, "was to link the spaces back to the wonderful Barbican itself and to celebrate the building's materiality by exposing the original concrete ceilings, de-cladding the hammered aggregrate walls in the Barbican Lounge space and using Cradley brick pavers for the Foodhall flooring.

Our second major direction was to create visual connections between the two offers, particularly via the outdoor terraces, which extend from each space. This was achieved through planting, but mostly by the design of huge, eye-catching and bespoke-designed umbrellas, made of two off-centre perforated aluminium disks, with the bottom disk measuring three meters in width, set into wooden bases, which house both planting and integrated seating.

Finally, we sought to animate the spaces with striking feature areas, details and materials, including a peacock-green resin floor in the Lounge, especially colour-matched to a photo, plus great lighting and an unusual mix of furniture. The new identity for the Foodhall uses

black and white photographic building icons and seeks to sit within the current Barbican brand family, although uniquely differentiated by the use of capital rather than traditional lower case lettering. "
Point of sale information in the Foodhall features handmade and bespoke steelwork slots, based on the information card slots on the front of Victorian architectural plan chests and a rolling blackboard used as a daily noticeboard. The Barbican Foodhall, is both a restaurant and a food market. Food producers represented in the space include Olives et Al, Severn & Wye Smokery, Hope & Green and Monmouth Coffee. For those who want to stay longer, the 200-cover restaurant features a new menu every day. The 150-cover Barbican Lounge on the first floor which offers small plate menus, gourmet bar snacks and a special Dine & Dash menu, where diners can be out of the door in 50 minutes. The Lounge also features London's first Macaroon Mixologist, where the legendary French delicacies are twinned with a range of new and traditional cocktails.

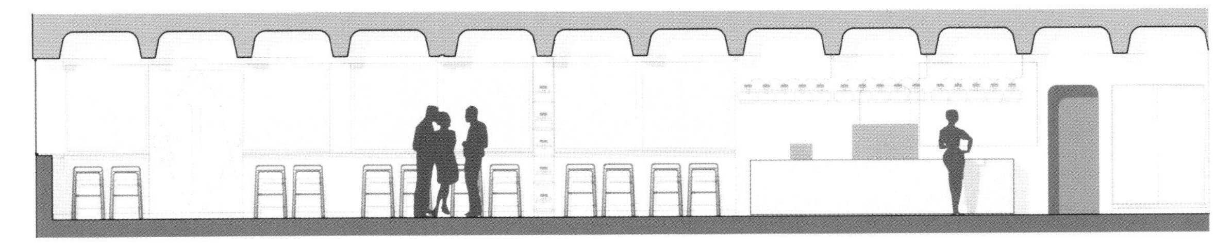

GROUND FLOOR

FIRST FLOOR

no smoking

If i loved you, i would tell you this
GREG MORTENSON
STONES INTO SCHOOLS
COLM TÓIBÍN
Brooklyn
BARBARA VINE
DANIEL GOLEMAN
ECOLOGICAL INTELLIGENCE
LITTLE GODS
ANNA RICHARDS
PICADOR
HOW TO SURVIVE THE END OF THE WORLD AS WE KNOW IT
JAMES WESLEY, RAWLES
FROM THERE TO HERE
NEENAW
SUZI BRENT
THE DAY OF THE TRIFFIDS
The Monday Night Cooking School
A Voyage Round John Mortimer
VALERIE GROVE
MEDIC
MEDIC
PAUL THEROUX
ANDREW RAWNSLEY
SERVANTS OF THE PEOPLE
TEN LITTLE HERRINGS
L.C. TYLER
Jonathan COE

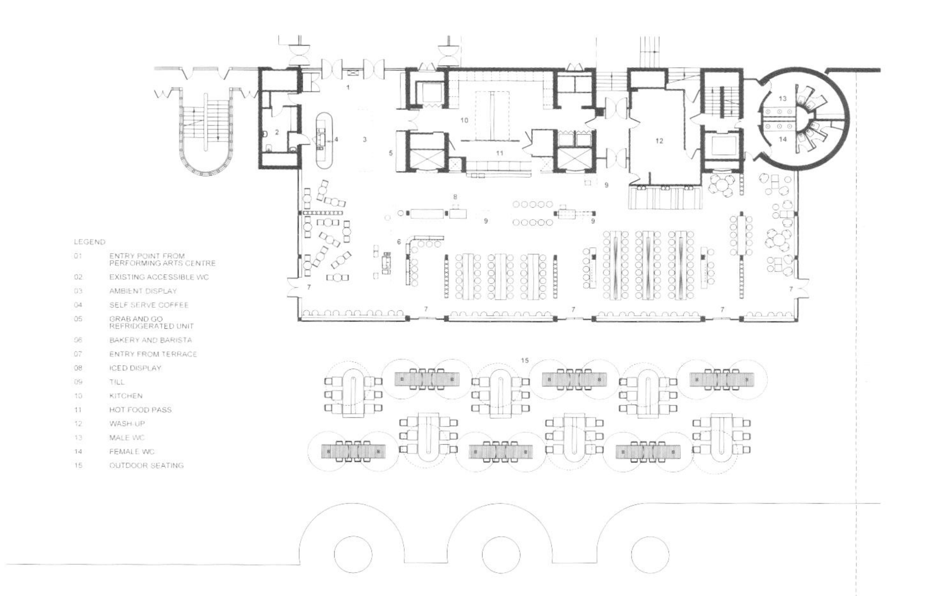
LEGEND
01 ENTRY POINT FROM PERFORMING ARTS CENTRE
02 EXISTING ACCESSIBLE WC
03 AMBIENT DISPLAY
04 SELF SERVE COFFEE
05 GRAB AND GO REFRIDGERATED UNIT
06 BAKERY AND BARISTA
07 ENTRY FROM TERRACE
08 ICED DISPLAY
09 TILL
10 KITCHEN
11 HOT FOOD PASS
12 WASH-UP
13 MALE WC
14 FEMALE WC
15 OUTDOOR SEATING

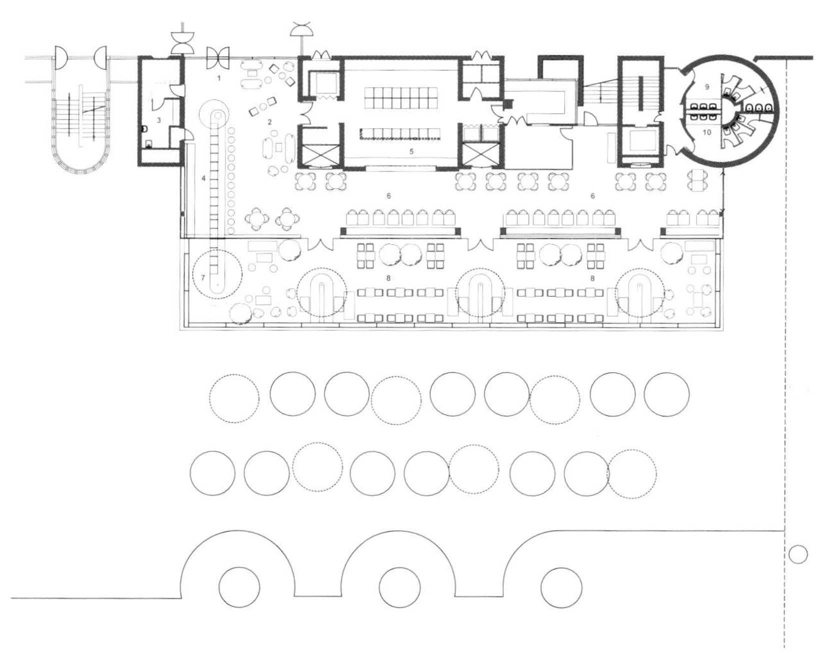

LEGEND

01 ENTRY POINT
02 LOUNGE SEATING
03 EXISTING ACCESSIBLE WC
04 BAR
05 THEATRE KITCHEN AND FOOD PASS
06 RESTAURANT SEATING
07 OUTDOOR BAR
08 OUTDOOR LOUNGE
09 MALE WC
10 FEMALE WC

SHH同时担任该项目室内设计以及品牌设计的工作。此处原址是450平方米的地下咖啡厅，现在改建成了餐厅和商店，重新命名为“巴比肯餐厅”。一楼现在包括“巴比肯休息室”以及室外阳台。设计师Helen Hughes解释道，“全局观”理念是着眼全局，从而将各个空间本身的精彩充分体现出来，而且凸显出建筑的原始性，例如，未经装饰的混凝土天花板、休息大厅里不经粉刷的水泥墙，以及由Cradley砖块铺设而成的地板。

第二个主要目标是将每个空间里延伸出来的阳台连接起来，从而在视觉上给人一种连接感。这一目标最终通过绿化得以实现，但大部分还是由巨大的特制遮阳伞来实现的。此伞是由两个中心穿孔的铝盘制成，底部的铝盘宽3米，伞下是客人的木质座椅以及绿树。

最后，力求以醒目的特色、缜密的细节和装饰材料来使整个空间充满活力。这包括休息室的孔雀绿树脂地板、色彩丰富的照片、巨大的照明灯和别具特色的家具组合。餐厅的新品牌标志使用了黑白色的建筑图标，尽管使用了大写字母来代替传统中的小写字母，此标志仍然与巴比肯家族的品牌风格保持一致。

餐厅用一个滚动的屏幕作为日常使用的布告牌，并可在上面进行食物报价。巴比肯餐厅，既是餐厅也是食品市场。在这里销售的食品有Olives et Al, Severn & Wye Smokery, Hope & Green，以及Monmouth咖啡。餐厅每天都提供健康饮食。在一楼的巴比肯休息大厅，每天都提供一些小吃、餐厅特制小吃以及特别的Dine & Dash。在这里，用不到50分钟，客人就可以尽享美味，满意而归。在休息大厅还有英国一流的调酒师，据传说，法国美食是由一系列传统的鸡尾酒孕育而成的。

DUTCH KITCHEN & BAR

Designers: Simone Pullens, Lieve Vandeweert
Design Company: Creneau International
Location: Amsterdam, Dutch
Area: 615 m^2

The renovated Boulevard at Amsterdam's Schiphol Airport can now help passengers really discover what Dutch is all about.

"As a hospitality organisation, we are always trying to make the passenger's day more cheerful," says Dawn Wilding, Vice President, HMSHost Europe and Middle East. "Guests visiting the new Dutch Boulevard will find innovative and interactive exhibitions, shops and places to eat. In the midst of Dutch Boulevard stands Dutch Kitchen Bar & Cocktails, the flagship of eating and drinking that reveals how HMSHost can create inspiring airport restaurants where passengers can enjoy the finest flavours from Dutch.

Interior architect Simone Pullens says: "Dutch may well be a very small country, but the Dutch are dreamers by nature and like to push back boundaries. Whether in technology, design, fashion or architecture, they often manage to outdo themselves. So the concept was self-evident - where a small country can do great things. When the idea of creating a Dutch setting in surrealistic surroundings. Typical Dutch elements , for example, tulips are enlarged enormously out of proportion, while a view of a Dutch polder landscape on the floor from above is minimized tremendously. The original wall tiles and the service sets are painted by a Royal Delft master painter."

The Dutch Kitchen and Dutch Bar are two different worlds that subtly merge thanks to a chequered floor pattern. The two worlds reflect the characteristic image of Dutch: classic on the one side, with houses on the canals and tulips; modern on the other side, with contemporary architecture.

Interior architect Lieve Vandeweert says: "We opted for expressive colours and materials to give visitors a real feeling of two worlds. The Dutch Kitchen exudes more charm, with light, bright colors, whereas the Dutch bar has a rather cosmopolitan and adult look, with mysterious, dark hues and gleaming textures."

DUTCH KITCHEN BAR&COCKTAILS

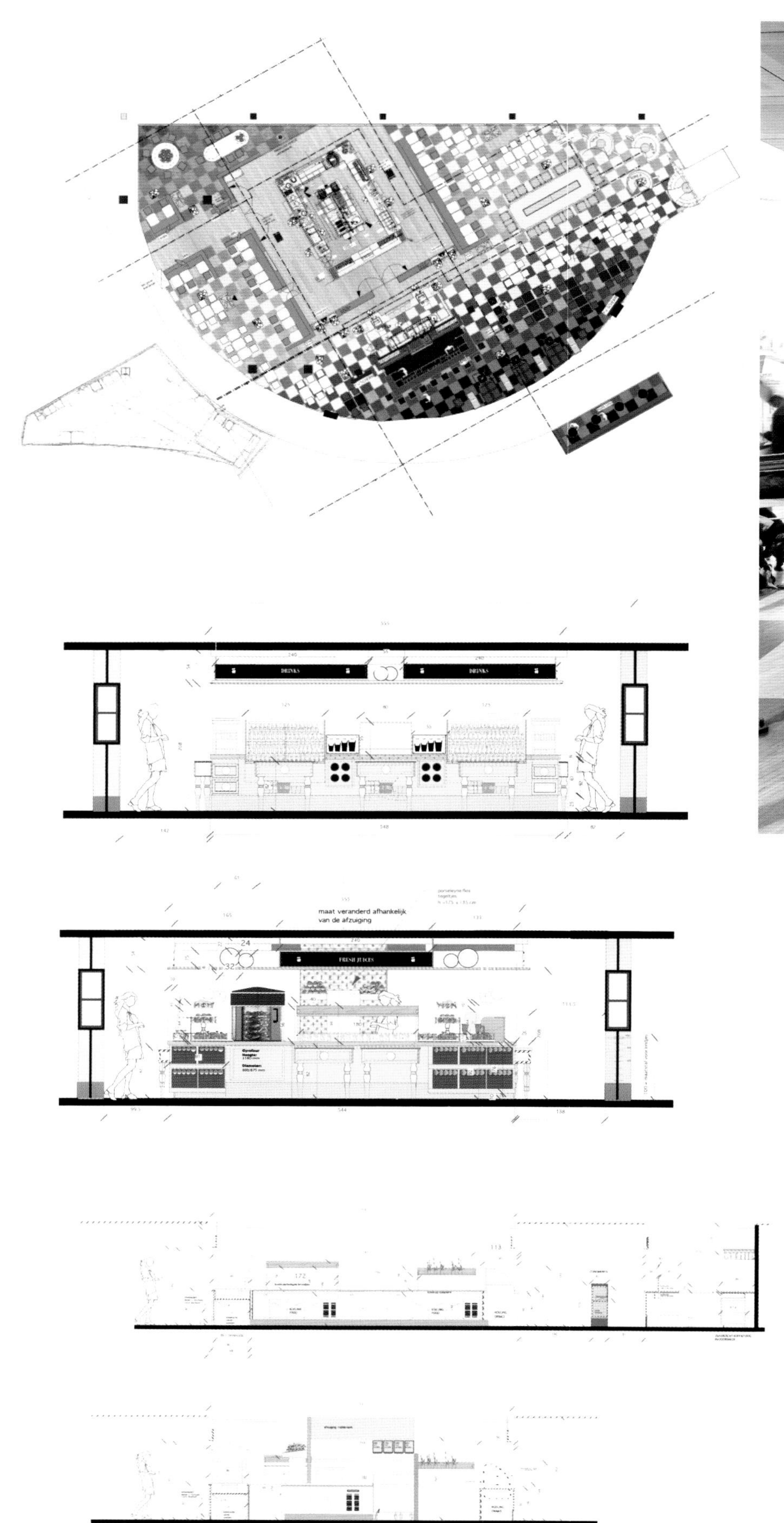

位于阿姆斯特丹史基浦机场的Dutch Boulevard，现在经过翻新，有助于客人全面了解荷兰。

HMSHost欧洲和中东区的副总裁Dawn Wilding说："作为一个热情的接待机构，我们努力让客人的每一天都过得更加快乐。""客人来到新Dutch Boulevard，会发现一些有创意和能交互的展览区、商店和就餐的地方。在Holland Boulevard的中央是Dutch Kitchen Bar & Cocktails，一个餐饮的旗舰店，展示HMSHost是如何创建鼓舞人心的机，给客人提供荷兰最好的美食享受。"

室内设计师Simone Pullens说："荷兰是一个非常小的国家，但是荷兰人天生就是梦想家，喜欢超越梦想。无论是在技术、设计、时尚还是建筑领域，他们总是设法超越自己。所以餐厅的设计理念非常清楚——一个小国家可以做伟大的事情。当有了在超现实主义的环境中创建一个荷兰背景餐厅的想法的时候，典型的荷兰元素被不成比例地进行放大，例如，郁金香；但是地面上荷兰的低洼开拓地景色，从上向下看，被无限地缩小。原来的墙砖和服务背景墙换成了皇家代尔夫特蓝（Royal Delft）的大师之作。"

"荷兰厨房"和"荷兰酒吧"是两个完全不同的世界，巧妙地融合了棋盘式的地板图案。这两个世界体现出了荷兰的特色：一方面很经典，房子建在运河和郁金香之上；另一面很现代，有很多当代的建筑。

室内设计师 Lieve Vandeweert说："我们选择有表现力的颜色和材料，令客人对这两个世界有真实的体验。"荷兰厨房"运用灯饰装饰和明快的颜色，充满魅力。"荷兰酒吧"则运用神秘、深邃的色调和闪光的质地，凸显国际化和成熟的外观。"

DUTCH
BAR&
COCKTAILS

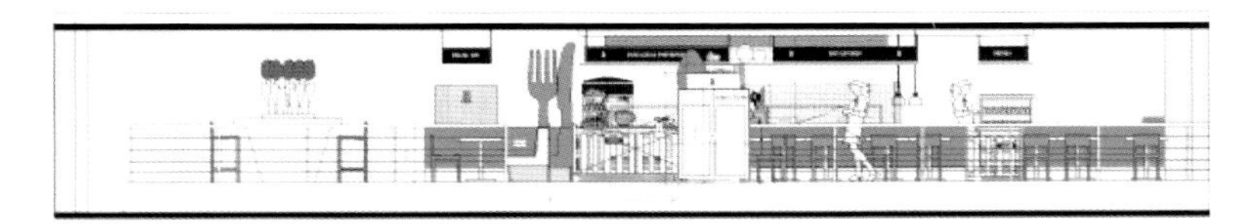

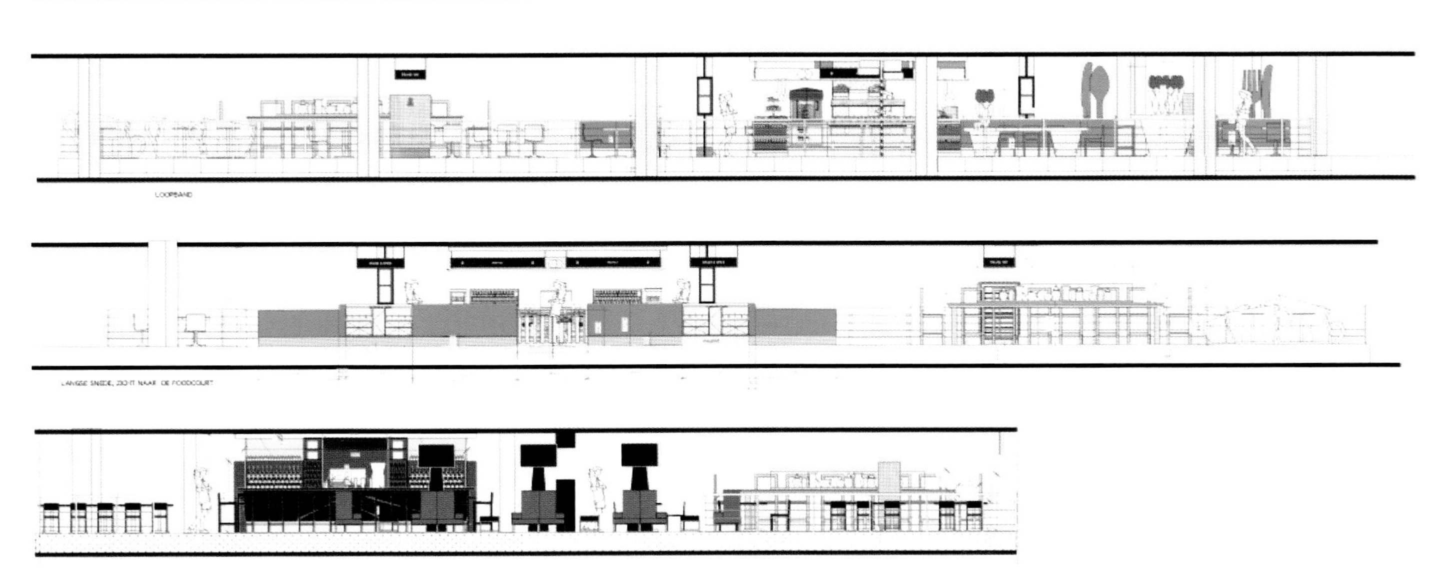
LOOPBAND

ROYAL DELFT 1653

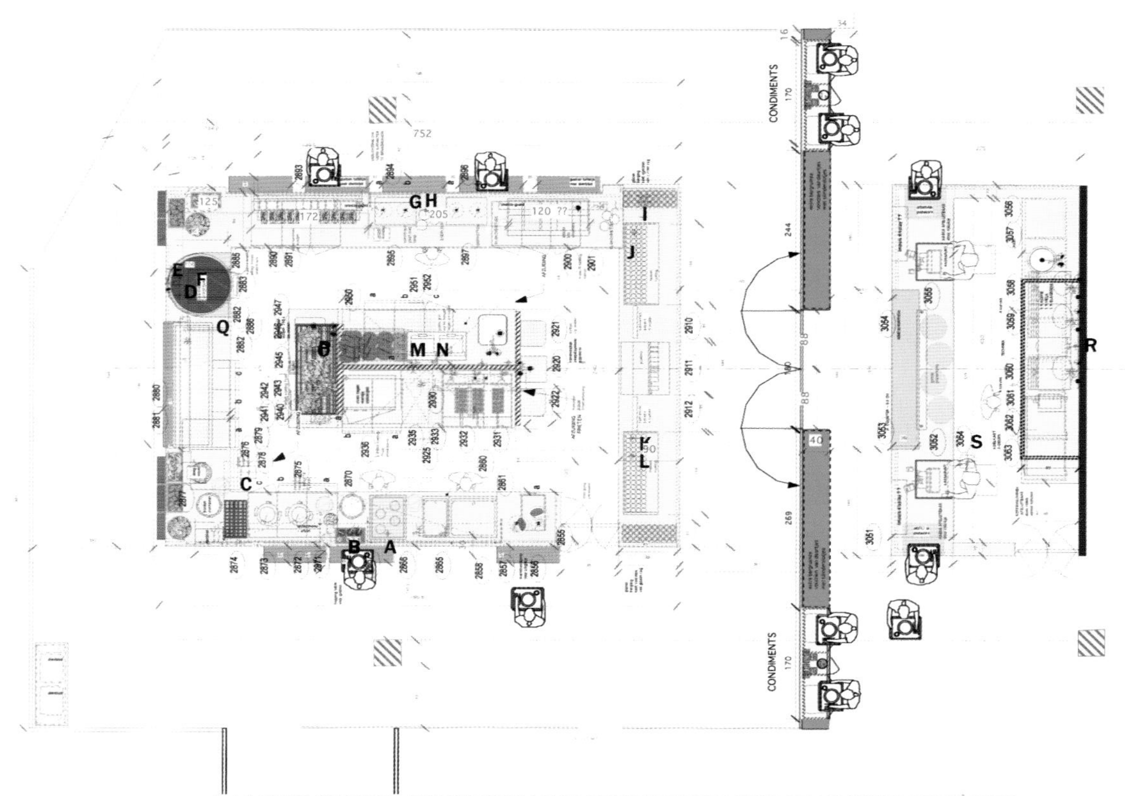
CONDIMENTS
CONDIMENTS
ALLE MAATVOERINGEN DIENEN TER PLAATSE GECONTROLEERD DOOR DE AANNEMER DER WERKEN

DUTCH KITCHEN BAR&COCKTAILS
BOLS COCKTAILS

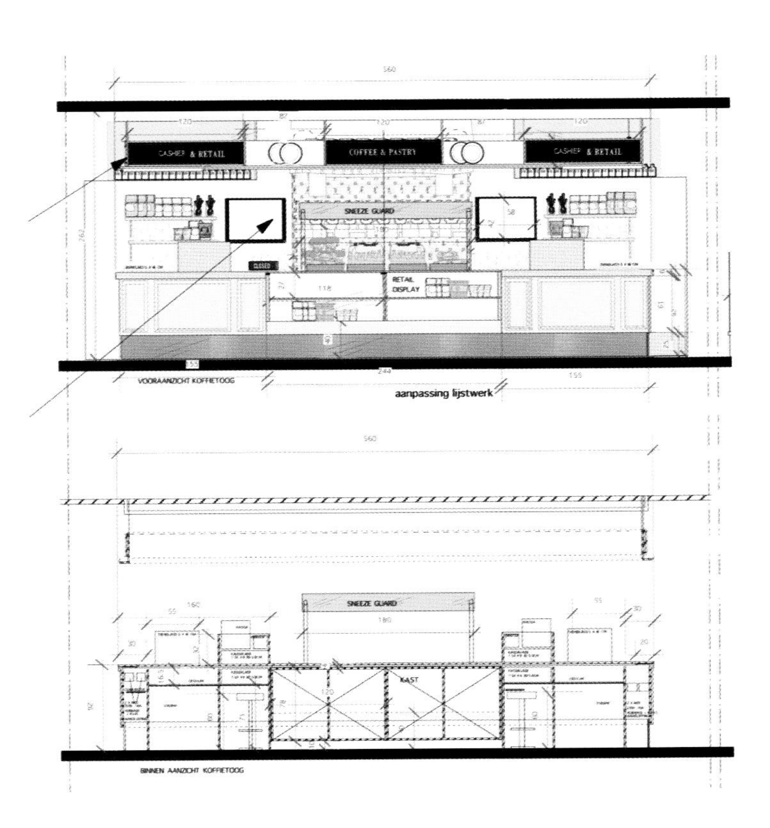
CASHIER & RETAIL
COFFEE & PASTRY
CASHIER & RETAIL
SNEEZE GUARD
CLOSED
RETAIL DISPLAY
VOORAANZICHT KOFFIETOOG
aanpassing lijstwerk
SNEEZE GUARD
KAST
BINNEN AANZICHT KOFFIETOOG

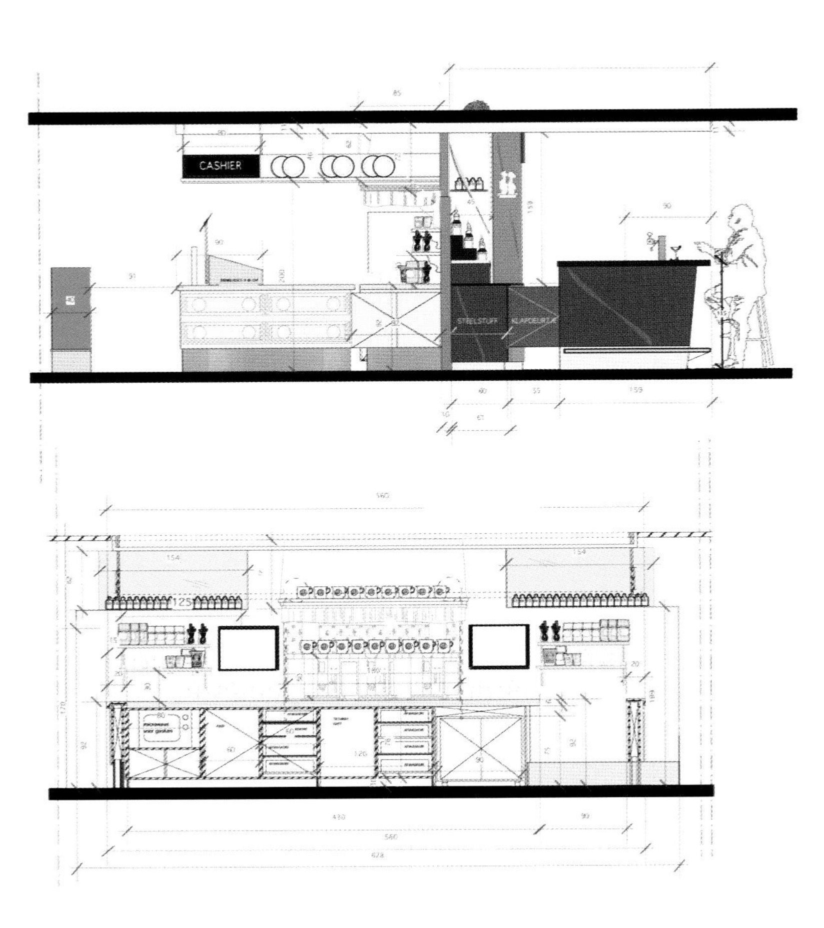
CASHIER
STEELSTUFF

CAVE RESTAURANT

Designer: Koichi Takada
Design Company: Koichi Takada Architects
Location: Sydney, Australia
Area: 104 m²
Photographer: Sharrin Rees

The designer wants to change the way guests eat and chat in restaurants. The acoustic quality of restaurants contributes to the comfort and enjoyment of a dining experience. The designer experimented with noise levels in relation to the comfort of dining and the cave-like ambience.

The timber profiles generate a sound studio atmosphere, and a pleasant 'noise' of dining conversation, offering a more intimate experience as well as a visually interesting and complex surrounding. The series of acoustic curvatures were tested and developed with computer modelling and each "timber grain" profile has been translated and cut from computer-generated 3-D data, using Computer Numerical Control (CNC) Technology.

Ever since the opening of CAVE restaurant, the success of the design has been judged by the overwhelming number of people who have come to eat and enjoy this unique dining experience. The CAVE restaurant is more than a simple interior, which has become a "place" of familiar identity, an address among the locals which offers an escape into "nature" from the urban surroundings.

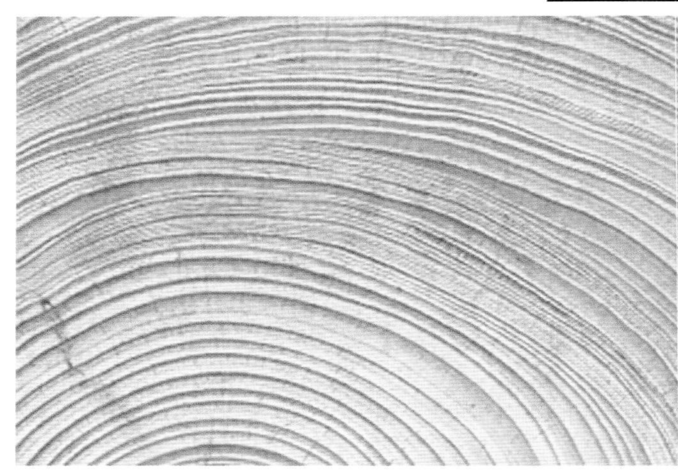

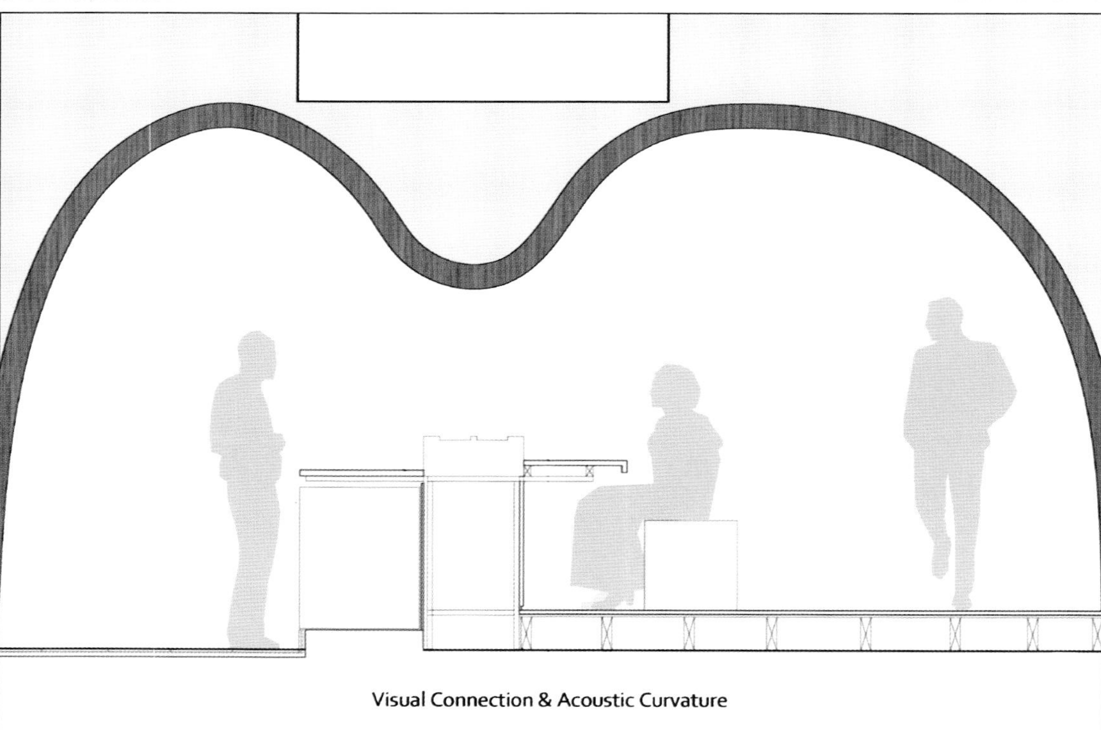

Visual Connection & Acoustic Curvature

设计师想改变客人在餐厅就餐和闲聊的方式。餐厅的口碑取决于客人感受到的舒适度与满意度。设计师做过与就餐舒适度有关的噪音标准实验和洞穴式就餐环境才有的氛围实验。

实木建材的外围产生了一种录音室的感觉，即使在此边聊边吃也不再扰人，反而让人觉得更加亲密，尤其是在这样一个极具视觉冲击力的复杂的环境中。弯曲的原生实木板材通过机控模具的制模和加工，每条树木纹理通过运用计算机数字控制技术3D数据进行分析和切分。

CAVE餐厅自开业以来，这种成功的设计理念已经被势不可挡的喜欢这种独特就餐体验的人们所证实。CAVE不仅仅是一家室内餐厅，它已然成了一个人们熟知的地方的象征，在众多饭店中脱颖而出，成为人们逃避喧嚣城市生活、亲近自然的好地方。

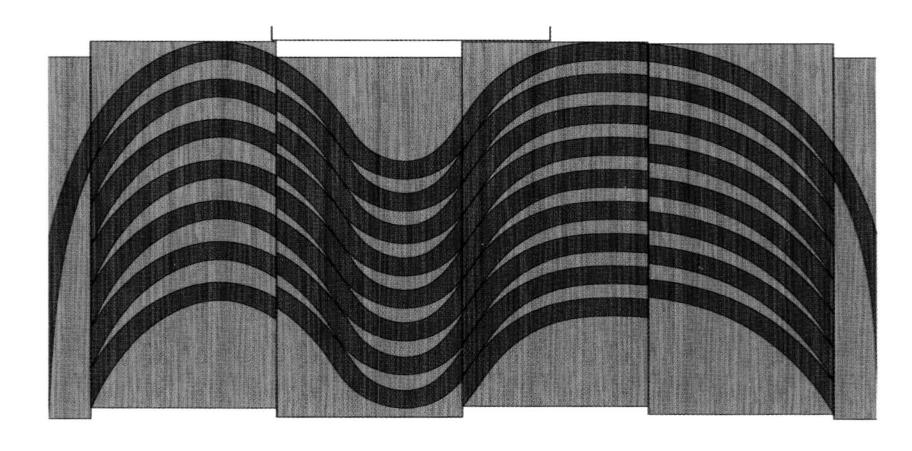

1.2 x 2.4m Acoustic Timber Set Out

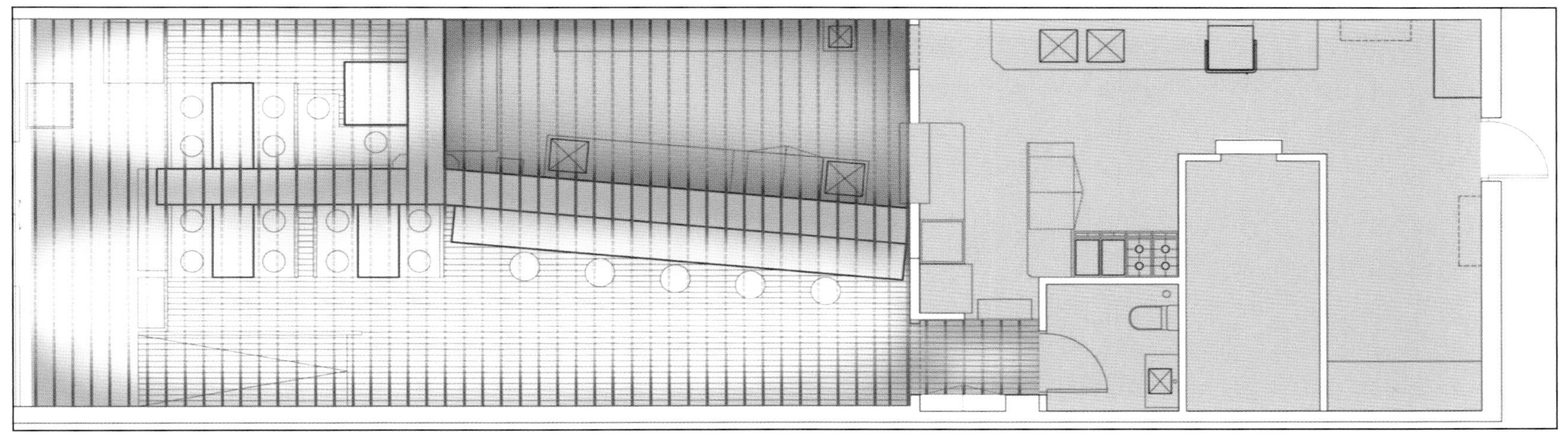

Acoustic Timber Ceiling Plan

1.2 x 2.4m Acoustic Timber Profiles

HOOKAH LOUNGE SATELITE

Designers: Mario Gottfried, Adrián Aguilar, José Alberto Rodríguez, Javier González, Óscar Flores, David Sánchez & Héctor Hernández
Design Company: BNKR Arquitectura
Partners: Esteban Suarez (Founding Partner), Sebastian Suarez
Project leader: Adrián Aguilar & Javier González
Collaborators: Jorge Arteaga, Zaida Montañana, Santiago Becerra
MEPs: SEI
Electrical Installations: Eledé
Lighting: Noriega Iluminadores
Construction: Factor Eficiencia
Location: Satélite, Estado de México, Mexico
Area: 553 m^2
Photographers: Fabiola Menchelli, Zaida Montañana

The design premise behind the Hookah Lounge is a fusion of a traditional Arabic style with a contemporary look. The roof garden of the restaurant has tables between ponds and walls composed of vertical gardens. Umbrellas resemble palm trees to further accentuate the illusion of this urban oasis. The end result was a strange and eclectic orgy of visual extravagance.

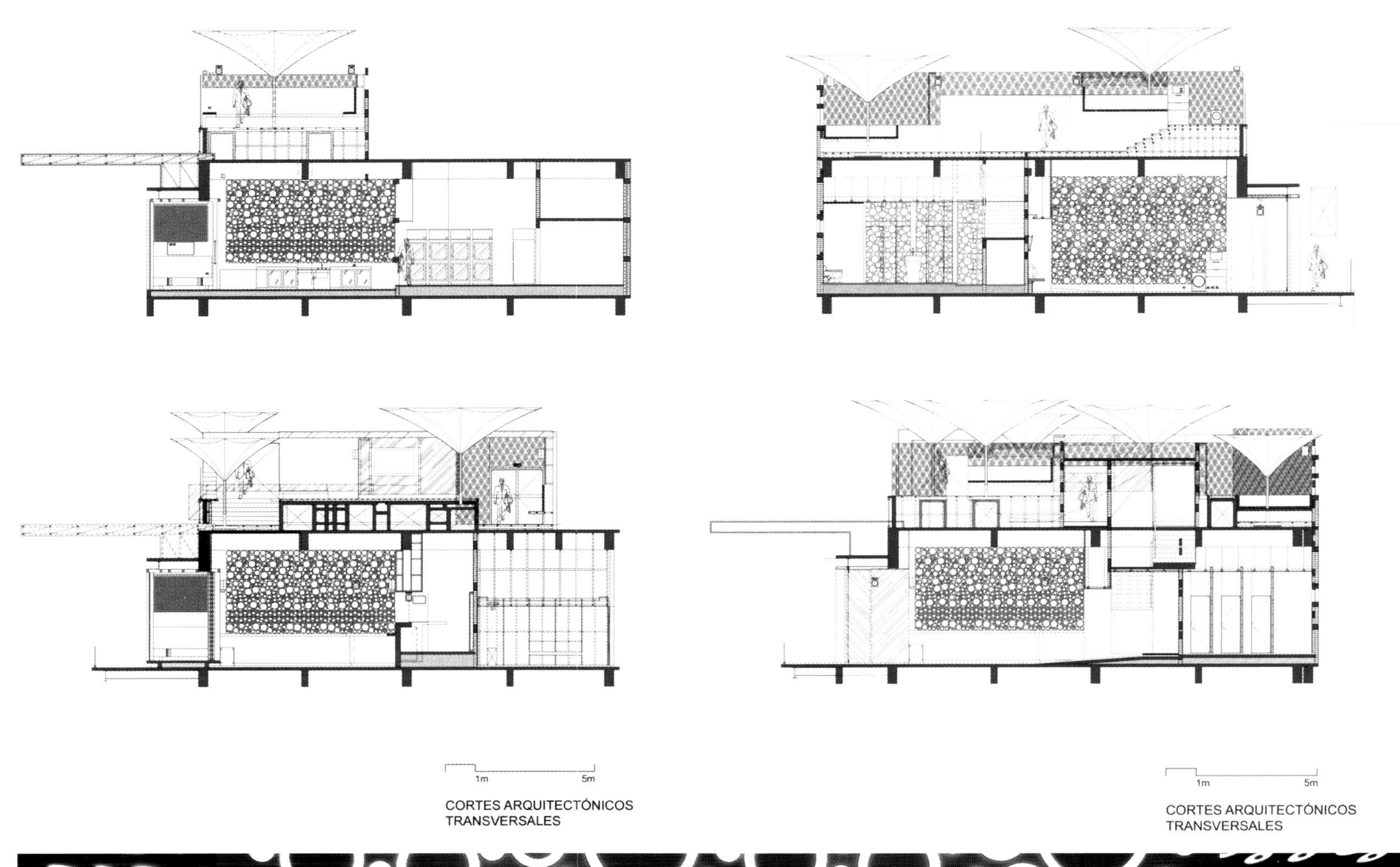

CORTES ARQUITECTÓNICOS TRANSVERSALES

CORTES ARQUITECTÓNICOS TRANSVERSALES

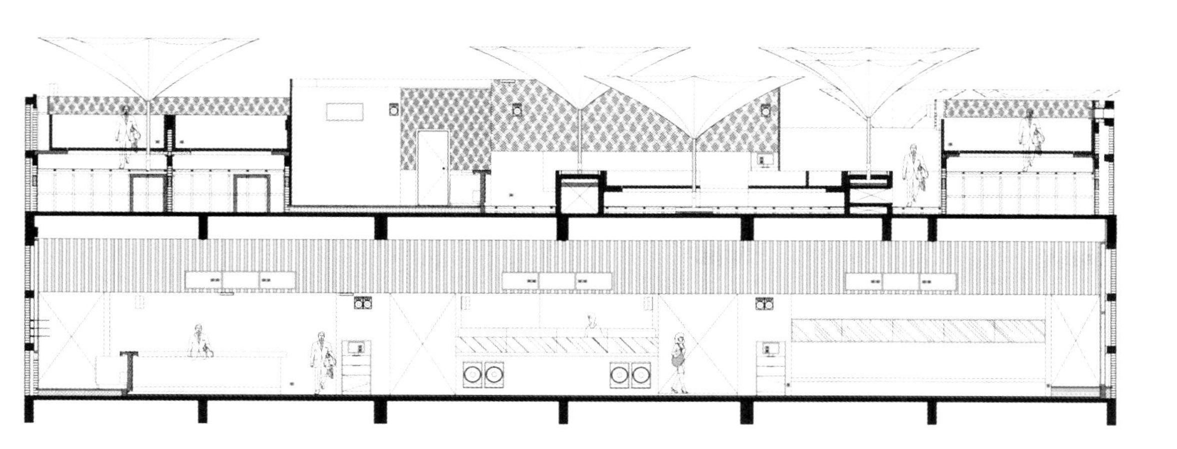

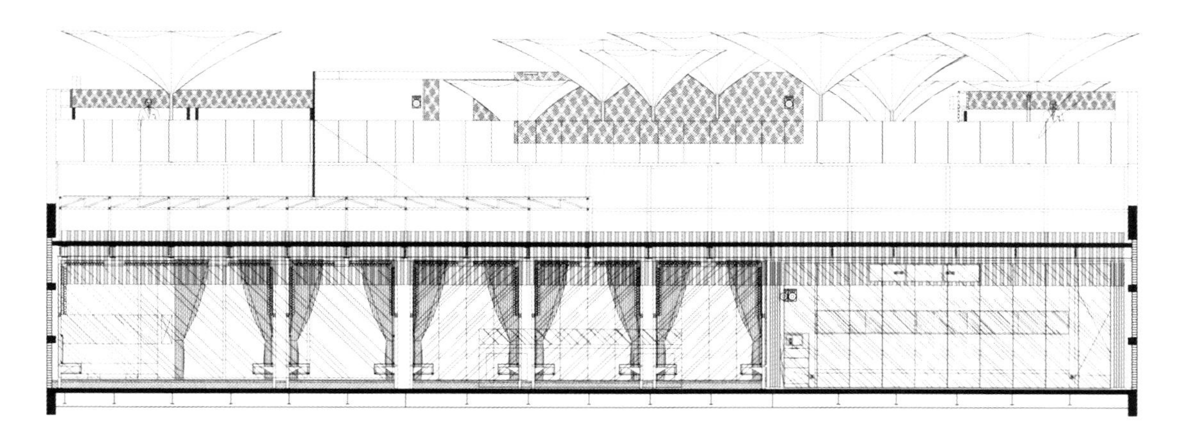

CORTES ARQUITECTÓNICOS
LONGITUDINALES

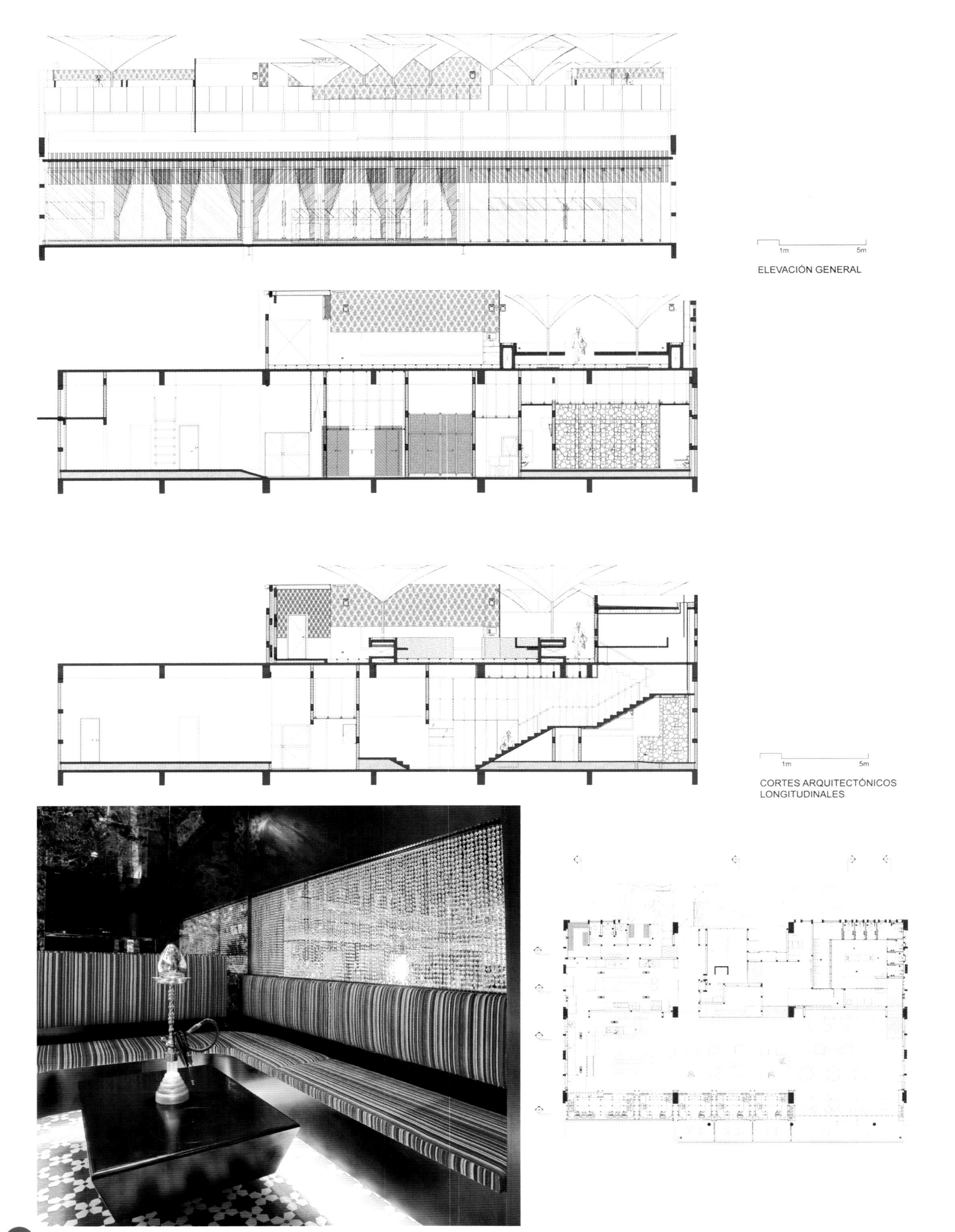

ELEVACIÓN GENERAL

CORTES ARQUITECTÓNICOS
LONGITUDINALES

PLANTA ARQUITECTÓNICA
CUARTO NIVEL

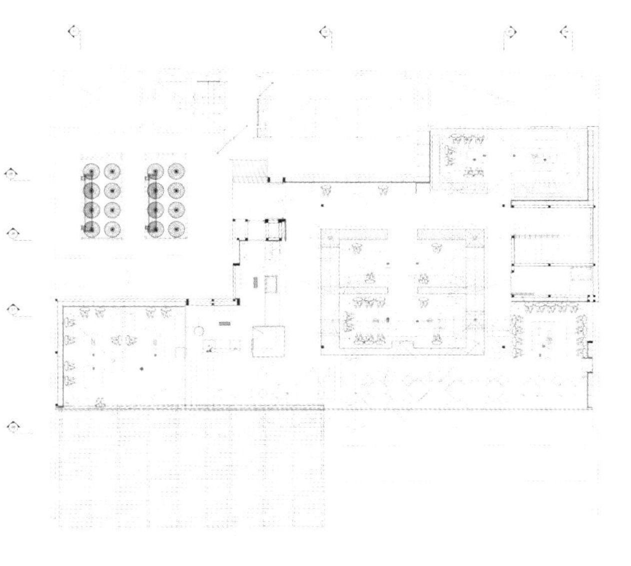

N
1m 5m
PLANTA ARQUITECTÓNICA
NIVEL AZOTEA

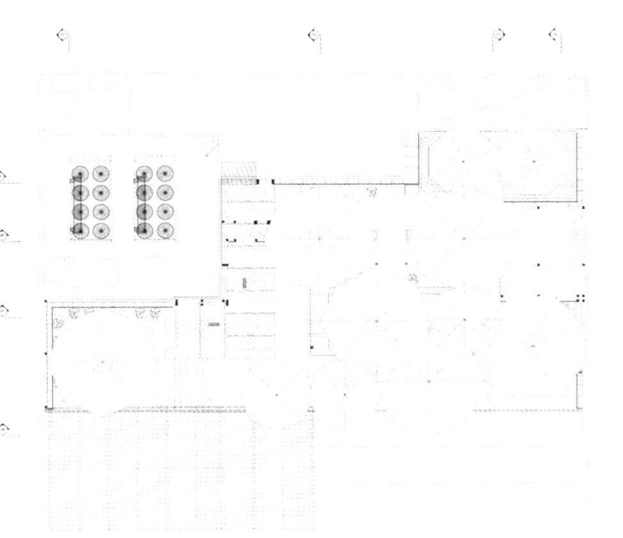

N
1m 5m
PLANTA ARQUITECTÓNICA
DE CUBIERTAS

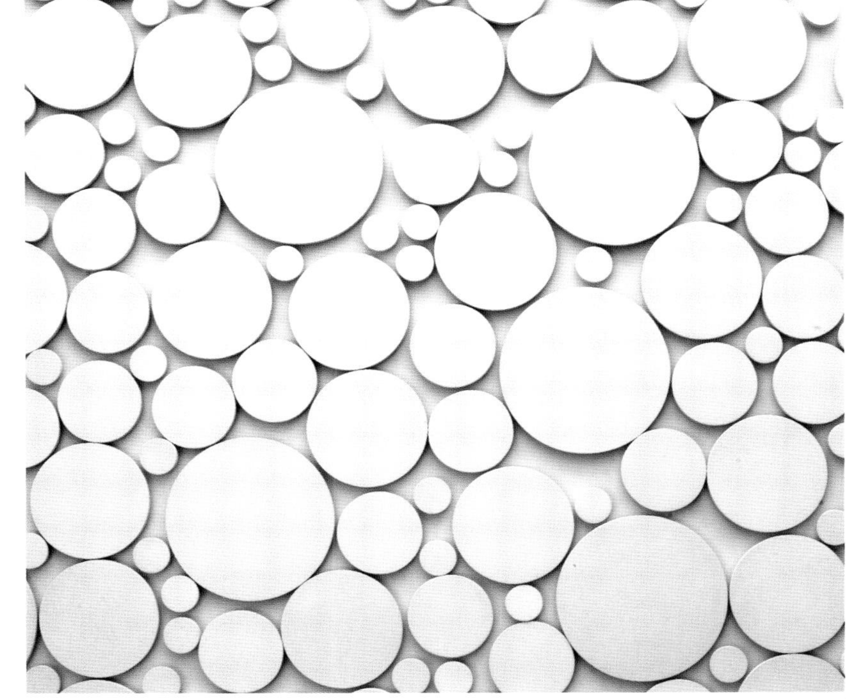

该餐厅建址于Hookah Lounge后面，兼具传统的阿拉伯风格和现代风格。餐厅的空中花园里以及池塘和立体花园之间都摆放了桌子。一把把遮阳伞看上去一像棵棵棕榈树，更加突出了城市绿洲的风格特色。设计作品最终带给客人一场奇特而又折中的、豪华炫目的视觉盛宴。

CAFE APRIL

Designers: Yuriy Ryntovt
Team: Andrey Gorozhankin, Maria Gorozhankin
Design Company: Ryntovt Design
Location: Moscow, Russia
Area: 600 m^2
Photographer: Andrey Avdeenko

A new frame-structured 5mx20m building was constructed in between two existing brick buildings. This allowed for the interiors to be open for viewing from the street.
Erected in 4 weeks the building was constructed with metal structures prepared beforehand.
Both sides of the facade were paneled with natural ash tree glued boards.
As a result of the construction of the new building and the reconstruction of the old one a floor was removed and a wholesome open space was obtained.
Additional function: cinema screen, stage, music and light controls, as well as conference room.
Project's idea: creation of conditions for small concerts, presentations, children's performances, and other informal communication.
Spacial structure: the restaurant was disposed on three levels, all open for public viewing as a single space. The lower level was equipped with a stage for public speeches and music performances.The ground or the middle level was where the main entrance and cloakroom were situated.The upper level had a conference room, a bar, and two restaurant halls.
Graphic design: photographic images of spring trees printed and applied on walls, curtains, textile and glass screens.

这是一个新的5米x20米的框架结构大楼，建造于两个现有的砖结构大楼之间，所以只能从大街上才能看到内部的景色。

这个大楼是在4个星期之内建造起来的，采用的是事先准备好的金属结构。

大楼的前后外墙都镶嵌了天然的白蜡树胶合板。因为既要改造旧楼又要新建大楼，所以原先的地板废弃掉了，还获得了开阔空间。

其他配置：播放电影的大屏幕、舞台、音乐和灯光控制、会议室。

项目理念：为下列活动打造出活动场地——小型音乐会、演讲、孩子的表演和其他非正式交流。

空间结构：餐厅共有三层，作为独立空间面向公众开放。地下一层打造了一个舞台，用于公共演讲和音乐表演。第一层（也就是中间一层）设置了入口处和衣帽间。第二层有一个会议室、酒吧和两个餐厅包间。

绘画设计：墙上、窗帘上、纺织物和玻璃屏幕上，都采用了春天里的树图像。

THE FOLLY

Designer: Sophie Douglas
Design Company: Fusion
Location: London, UK
Area: 1,230 m²

On the ground floor, the guest enters into the terrace area,with trees, a stone random mosaic floor and a real garden vibe.

From the terrace, turn right into the Potting Shed, housed in its own rough timber structure and lit with lamps. The open kitchen and bar, with their polished concrete front, run continuously along the back wall of the space. A wonderfully tactile natural cork bar, and once again, the back bar, formed of boxes covered in Liberty print style papers, is backed by the same, similarly lit, rough brick wall as the ground floor.At the kitchen, a dramatic and open operation, backed by sage green wall panels, customers are encouraged to sit and watch the chefs at work. Adjacent to the kitchen, the chefs table area, another individual room space, is wrapped in bold botanical wallpaper, offer the perfect environment for a luxurious lunch or dinner party.

The main restaurant area is furnished with a combination of long custom made trestle tables and brightly, but luxuriously upholstered benches, and tables for two and four, with character oak tops and fifties style arm chairs, upholstered with leaf and floral print and velvet fabrics.

At the opposite end of the bar, an eight-meter-high tree trunk emerges from the basement through the floor. And into the library area, high tables provide a great place to grab a magazine, a coffee and a pastry.

At the front of the space, to maximise the light, a large void in the floor joins the ground and lower level bars, and houses a dramatic open stair. Arriving in the centre of the lower space, one is faced with a wonderfully tactile natural cork bar, and once again, the back bar, formed of boxes covered in Liberty print style papers, is back by the same, similarly lit, rough brick wall as the ground floor. Opposite the bar, running along under the pavement, a series of raised vaults, with their own individual staircases, are wrapped with luxurious velvet banquette seating and the simple white walls are decorated with, large scale, intricate, botanical and floral artwork. A series of circular booths, topped with lime green anglepoise silk shades, sit at the centre of the space, against a back drop of a flowing, fret cut timber screen and the rough cedar cladding of the rear wall that houses a beautiful and intimate private dining room.

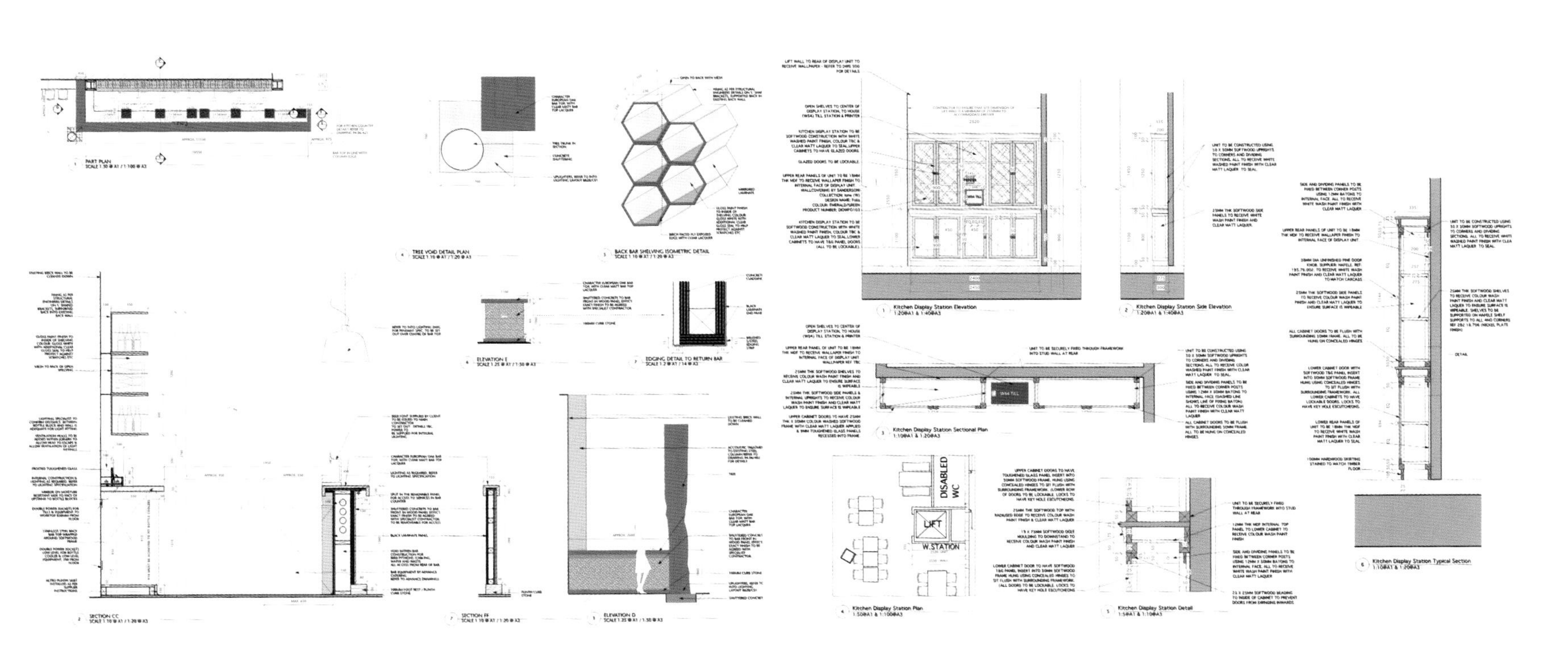
PART PLAN
TREE VOID DETAIL PLAN
BACK BAR SHELVING ISOMETRIC DETAIL
ELEVATION E
EDGING DETAIL TO RETURN BAR
SECTION CC
SECTION FF
ELEVATION D
Kitchen Display Station Elevation
Kitchen Display Station Side Elevation
Kitchen Display Station Sectional Plan
DISABLED WC
LIFT
W.STATION
Kitchen Display Station Plan
Kitchen Display Station Detail
Kitchen Display Station Typical Section

在一楼，客人可以进入到长满树木的露台区，这里有用石头随意砌成的马赛克地板和一处真正的花园。

从露台右转是盆栽区，由粗糙的木质结构建造而成。长长的开放式厨房和吧台沿着空间的后墙，吧台前面则采用混凝土打磨而成。吧台后侧是类似蜂巢的镜盒，打造在纹理粗糙、灯光照亮的砖墙上。厨房内，厨师神奇的操作在鼠尾草绿的背墙的衬托下尤为精彩，吸引客人前来观看。靠近厨房的是厨师餐桌区域，墙上贴着轮廓清楚的植物墙纸，这是另外一个独立的空间，为奢侈午餐或者晚宴提供了完美的环境。

餐厅的主要区域内摆放着定制的长支架式活面餐桌和颜色鲜亮、有点奢华的长凳，可供两人或四人就餐的桌子和橡木桌面，还有20世纪50年代风格的扶手椅子，椅子上铺着天鹅绒坐垫，坐垫上印着树叶和花朵的图案。

在吧台的对面方向是一个8米高的树干，从地下穿过地板长出来。在图书馆区，高高的凳子提供了一个很好的场所，可以供客人阅读杂志、喝杯咖啡、品尝小点心。

在餐厅的前面，为了充分利用灯光，在中央设置一块空间，很好地融合到餐厅的一楼和较低地势的吧台，中央还有一个视野开阔的楼梯。进入这个空间中央，能看到一个非常有质感的自然软木吧台，紧接着就是餐厅后面的吧台，由Liberty印花包装纸包装的盒子打造而成。后墙仍采用用近似灯光照亮的、粗糙的砖墙。吧台的对面，餐厅内人行道的下面，有一系列高度升起的拱顶区域，每个区域都有自己独立的楼梯，座位的表面是用奢华的天鹅绒做成的，后面的白墙也是简简单单，只用了大型的、神秘的植物和花的图案来装饰。圆形的雅座位于这个空间的中央，是由淡黄绿色丝绸做灯罩的安格泡灯（Anglepoise）进行照明。雅座的后面是万字浮雕装饰的木质板面，以及用杉木所包覆的后墙，整体营造出一个漂亮且舒适的私密就餐空间。

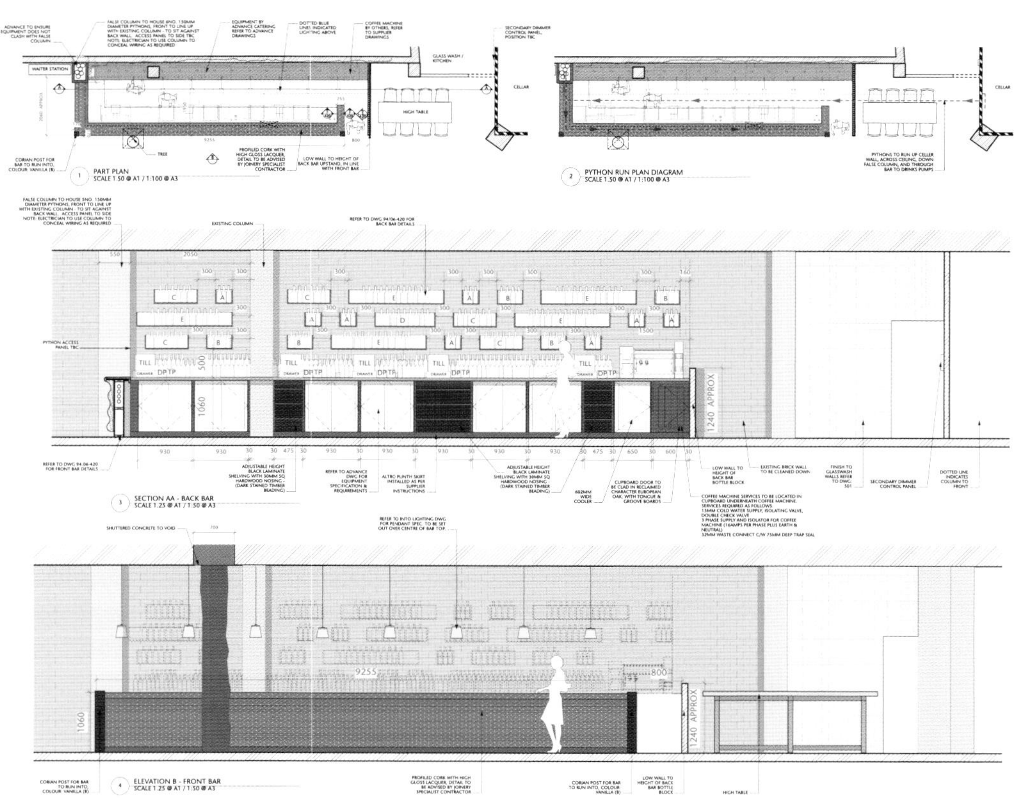
PART PLAN
SCALE 1:50 @ A1 / 1:100 @ A3
PYTHON RUN PLAN DIAGRAM
SCALE 1:50 @ A1 / 1:100 @ A3
SECTION AA - BACK BAR
SCALE 1:25 @ A1 / 1:50 @ A3
ELEVATION B - FRONT BAR
SCALE 1:25 @ A1 / 1:50 @ A3
1240 APPROX
1060
9255

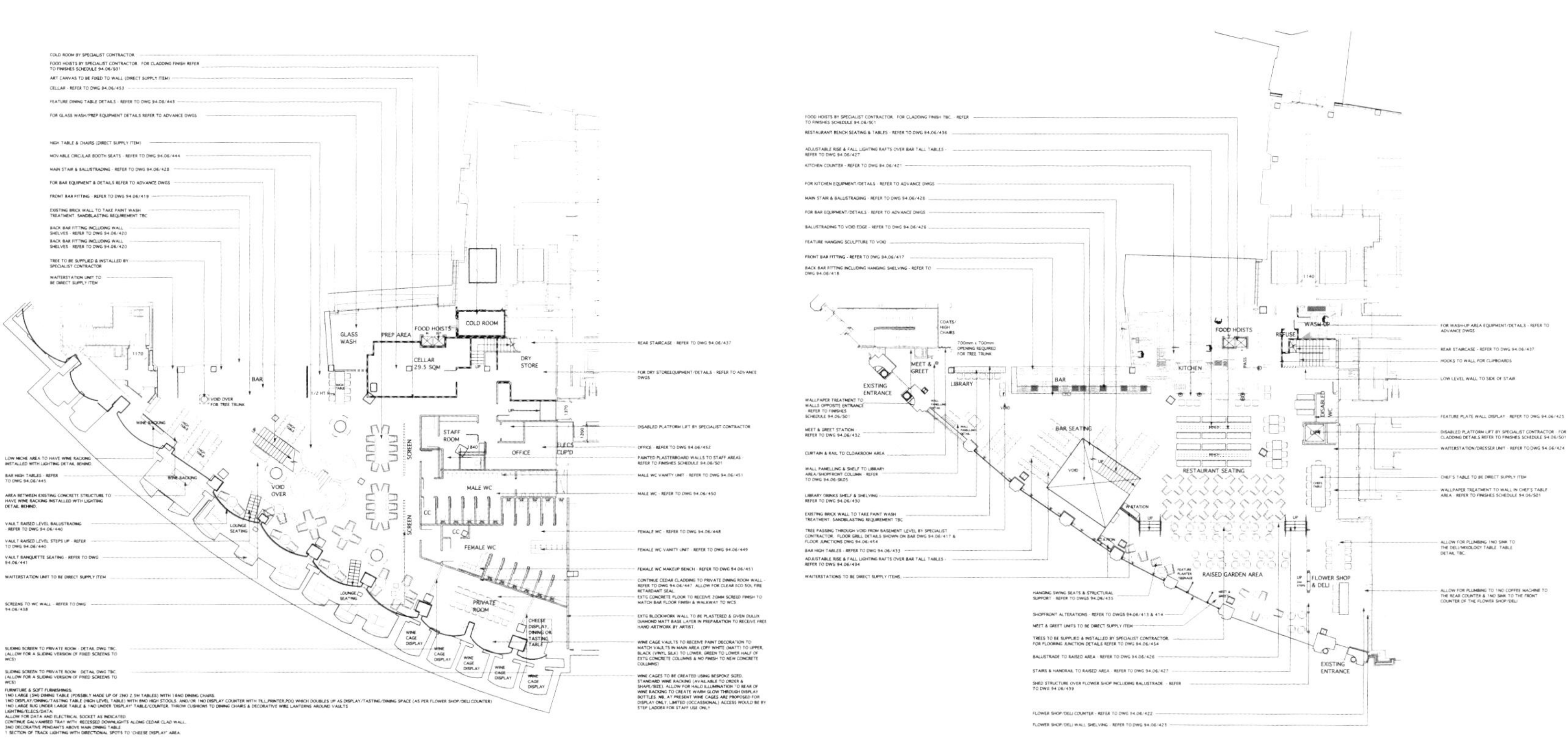

GLASS WASH
PREP AREA
FOOD HOISTS
COLD ROOM
CELLAR 29.5 SQM
DRY STORE
BAR
STAFF ROOM
OFFICE
MALE WC
FEMALE WC
SCREEN
PRIVATE ROOM
EXISTING ENTRANCE
MEET & GREET
LIBRARY
BAR SEATING
KITCHEN
RESTAURANT SEATING
RAISED GARDEN AREA
FLOWER SHOP & DELI

THE FAIRWAYS HOTEL AND SPA

Designer: Darley Interior Architectural Design (DIAD)
Location: Johannesburg, South Africa

DIAD, the hotel specialist interior design team, created a Contemporary-African look and feel to the interiors blending innovative ideas within well-designed spaces.
Overlooking the extensive views of the rolling green landscape of golf fairways and densely treed suburbs, DIAD's design inspiration for the concept was to weave this surrounding environment into the interiors within the strong contemporary background of the architecture.
With the hotel, being aimed at international Johannesburg visitors, weekend golfers, weekly conferences and day spa retreats, the interiors needed to reflect an international, contemporary elegance whilst providing a subtle sense of place.
The restaurant and bar venues where also required to entice and attract a loyal local following providing a trend-setting and timeless destination.
The DIAD team sought to reflect and integrate the surroundings by selecting natural finishes and designing features that could be manufactured by South-African craftsmen, utilizing environmentally friendly and recycled material and products.
The particularly successful element of the Interior Design of this project is the seamless integration of spaces throughout the hotel extending to and including the surrounding golfing greens. Driving through this Johannesburg suburb, the guest would never speculate that nestled comfortably and majestically alongside one of the most exclusive golf courses lies this oasis with its subtle Contemporary-African flavor.

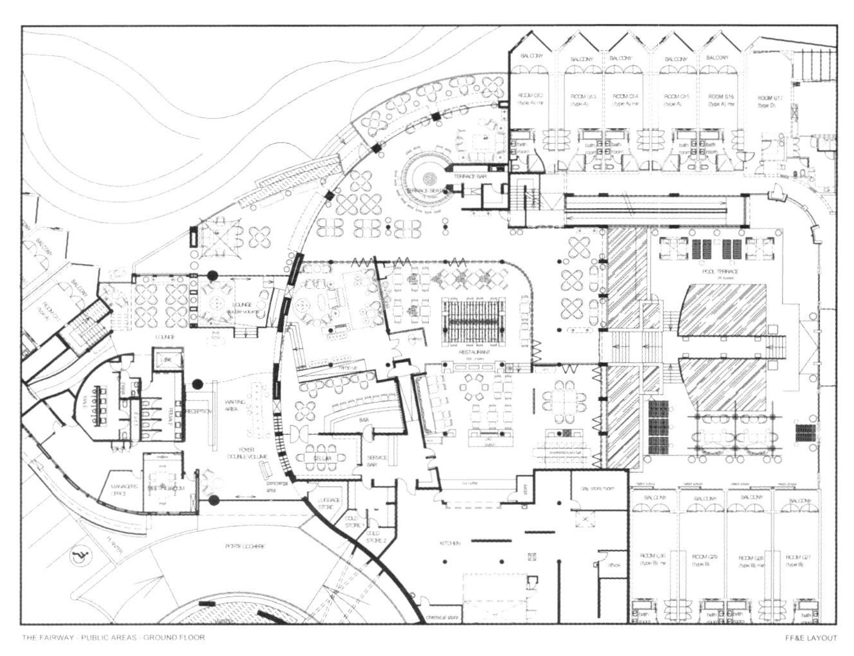

THE FAIRWAY - PUBLIC AREAS - GROUND FLOOR

FF&E LAYOUT

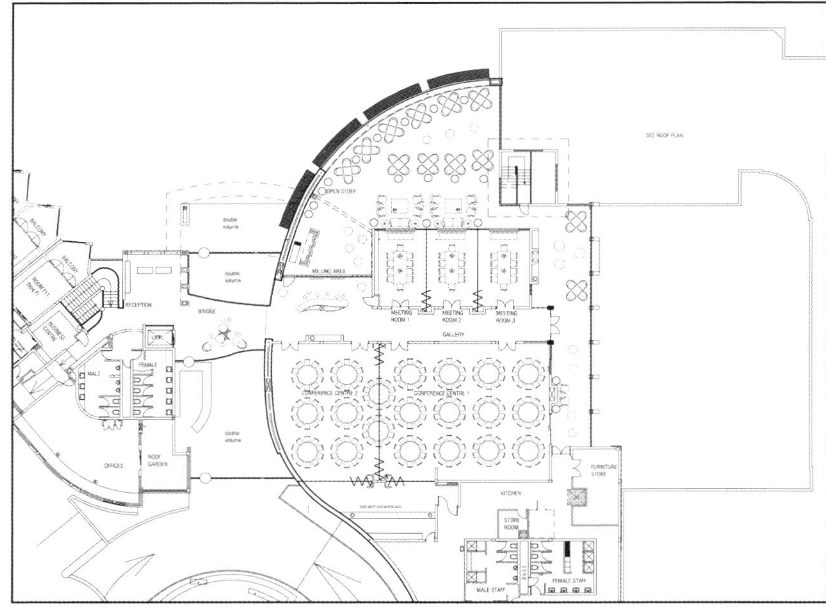

THE FAIRWAY - PUBLIC AREAS - FIRST FLOOR

FF&E LAYOUT

DIAD是酒店室内装修设计团队专家，融合了创新理念，营造了具有现代非洲气息的高尔夫酒店环境。

俯视绵延翠绿的高尔夫球道和绿树成荫的周边，DIAD的创作灵感是把这样的外景呈现在室内设计上，突出建筑的现代风格。

酒店的目标客人是来自约翰内斯堡的游人、周末来打高尔夫的人，以及来开周会的人和每天做Spa的人。这就需要把酒店打造成具有国际化需求、精致，同时又不失高雅的一个环境。

同样，就餐区和酒吧区也要吸引当地忠诚的客人来此休闲度假，并逐渐形成一种时尚。

DIAD团队选择天然元素和出自南非工匠之手的特色工艺，并使用环保材料和可回收再利用的建材，力求在作品中反映并融合自然环境特色。

该项目设计的点睛之笔是酒店的各个空间，包括高尔夫场上绿地，转呈自然，浑然一体，毫无分割断层之感。驱车行驶在约翰内斯堡城郊，绝对不会想到如此高级的高尔夫球场之边会有一个极具现代非洲气息的一个休闲场所，如同绿洲上点缀着的鸟巢，舒适又不失尊严。

WESTMINSTER BRIDGE PARK PLAZA

Designers: Eyal Shoan, Oded Hagai, Iris Rubinger, Eldad David Husravi, Liza Grishakova
Design Company: Digital Space
Area: 1,400 m^2

At ground floor level Digital Space located two large, LED lit glass feature walls, and positioned one on each side of the main staircase. This staircase leads to the first floor public areas , including the restaurant "Brasserie Joel", Ichi Sushi & Sashimi Bar and Primo Bar.
The "Brasserie Joel" restaurant serves as the hotel's guest's breakfast in the mornings and as grommet restaurant for lunches and dinners. For that reason, Digital-Space created versatile buffet area. The the restaurant was divided into two areas. They made it by huge artwork that used as semi transparent glass filter wall between the areas. Clean lines and luxurious materials, brought together with this artwork and variety of design features, ensure unique guest experience.
One of these features is a large wine cellar – created from bespoke acrylic shelves, curved to hold more than 450 bottles and set into a temperature controlled glass cabinet. The wine cellar lit by RGB LED light to match the colors of the artwork. Curved floating wood ceilings – which grow from the benches – define the seating areas and add visual interest to the space.
This intimate Japanese sushi restaurant is cool and contemporary in design with a choice of seating at the sushi bar, watching the skilled sushi chefs, or at tables throughout the restaurant.
The Ichi sushi bar is located in a close box, with great view to the Big-Ben. By using various materials – glass, wood, Japanese Washi and leather – the designers created modern space with traditional nature.
Soft wall behind the sushi counter is folded to be also the ceiling and gives floating feeling to the space. The envelope created from two layers – wood louvers on the front and glass on the back, with RGB light in the middle that emphasize the differences between the materials. On the back part, there is a big lit glass wall with Washi pattern, which add traditional ambiance to the space.
The Primo Bar itself is used to separate the spaces on the first floor, designated to serve as a lounge and bar making the bar a design feature in itself. Needing to create two different spaces, particularly in the evenings, the area on the external side of the bar features a stage where live acts can perform and a dance floor when tables and chairs are removed. On the other side of the bar, nearer to the main entrance, comfortable modern designed sofas and armchairs allow this lounge area to be a place to relax.

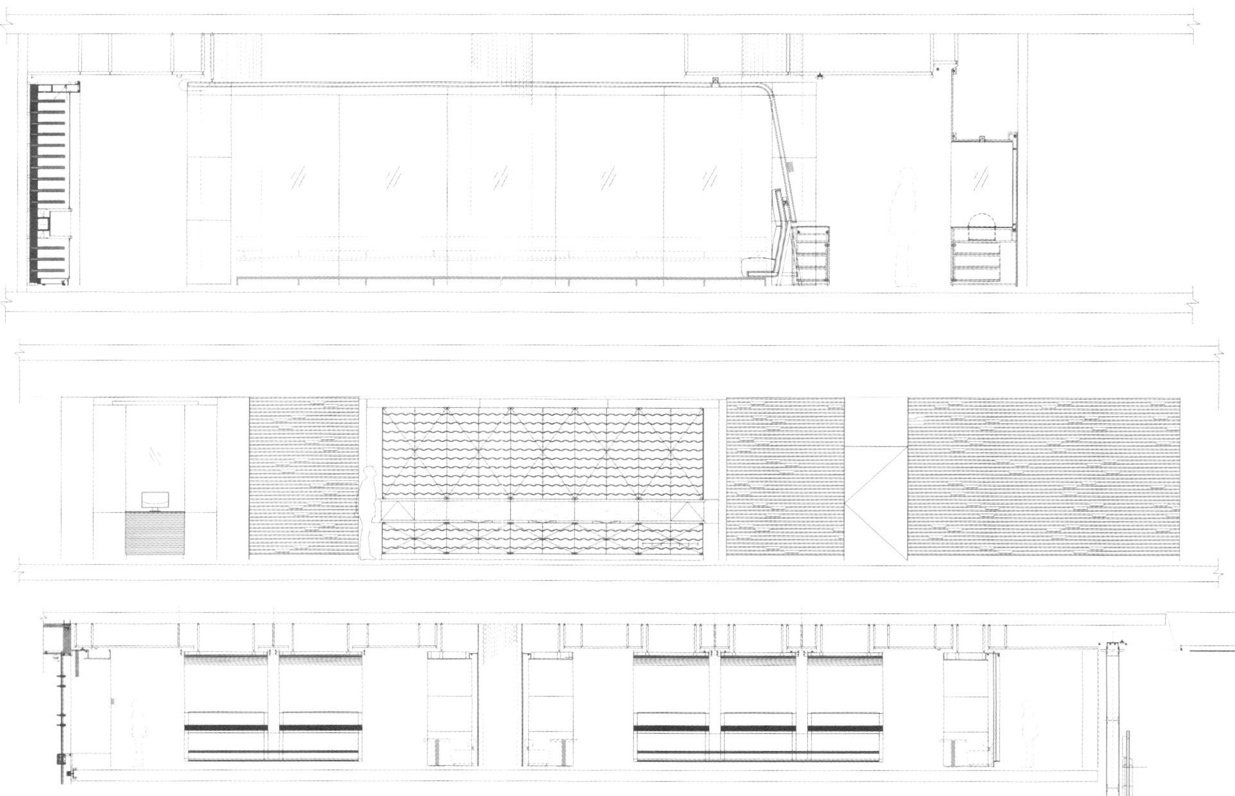

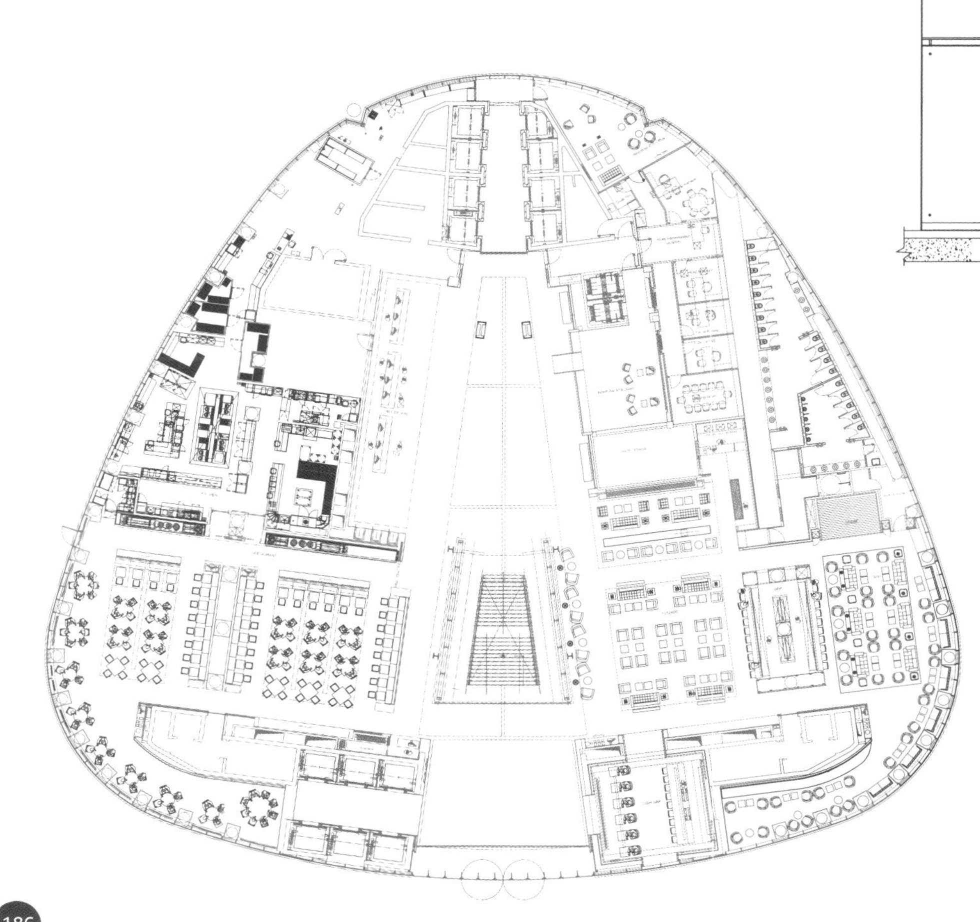

Ichi寿司店在一楼， Digital Space设计公司安装了两个巨大的LED灯射玻璃墙，分别矗立在主楼梯的两侧。楼梯一直通向一楼的公共空间——包括Brasserie Joel和Ichi Sushi & Sashimi Bar，以及Primo Bar在内的巨大空间。

Brasserie Joel餐厅主营早餐，兼供午餐和晚餐。因此，Digital Space需要设计出一个多功能就餐区。餐厅被分割成两部分，设计师用半透明的玻璃板打造出来的巨幅精美栏板将两部分隔离开来。洁净的线条、奢华的取材携手成就了这里的特色，确保客人能够享受到世上绝无仅有的美食体验。

宽敞的酒窖是这里的另一特色。专门定制的丙烯酸塑料构架弯曲而下，架构空隙里有不止450个酒瓶，固定在一个温控玻璃房下。在RGB LED灯照射下，酒窖与巨幅栏板交相辉映，相得益彰。

这家私密的日式寿司店的设计给人一种清凉而又现代的感觉，坐在寿司吧前，客人可以欣赏着师傅们的寿司技艺，抑或是坐在桌旁小憩。

寿司吧设在一个临近的包间内，落座其中，可以将大本钟尽收眼底。不同的建材——玻璃、实木、日本Washi和皮革——设计师借此打造了一个具有传统文化底蕴的现代空间。

寿司吧台后面的柔面墙可以折叠起来充当天花，空间显得流畅而自然。两个楼层间的夹层——实木天窗在前，玻璃天窗在后，中间的RGB灯使得两种建材对比更加强烈。后面巨大的灯饰玻璃墙上的Washi花纹，将传统的格调融入其中。

Primo Bar本身就是一楼的一个隔间，主要功能就是回廊和酒吧。为了营造两个风格迥异的空间，特别是在夜晚，这个空间的外围设计需要突出其“舞台”功能，挪开桌椅，这里就可以用来演出或作为舞厅使用。酒吧的另一侧，距离主入口较近，舒适、现代的沙发和扶手椅吸引客人在这里怡然小憩。

Resorts

YOSHIZUSHI

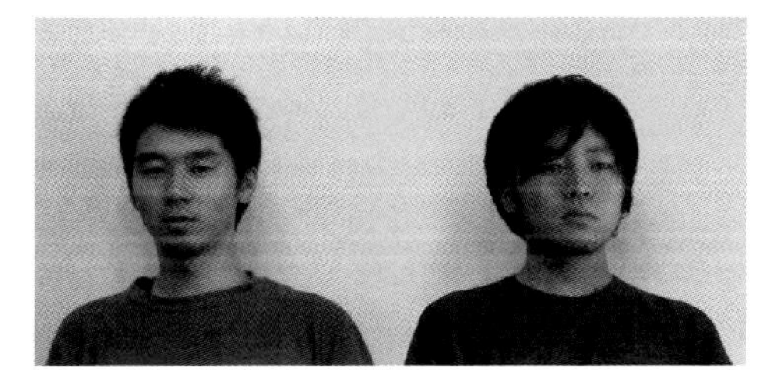

Designers: Naoki Horiike, Norihisa Asanuma
Design Company: Epitaph
Location: Meguro-ku, Tokyo-to, Japan
Area: 50 m^2
Photographer: Mitsunobu Horiike

This project is remodeling of the sushi shop in Japan.
This store which was in the first floor of the building was redecorated with remodeling construction of the whole building and opened in October, 2010.
The thing that the designers demanded was thing that work of cock can be efficiently done and thing to make the inside of a store full of bright and simple space. The designers noted that a simple and clean impression of the sushi was not ruined.
In addition, the designers did the operation to change the color of a wall and the ceiling of the kitchen into to create holy areas of the chef. A ceiling is low under the influence of the existing skeleton, but does not feel it because it is the simple plan that the whole can look around.
The counter of the Japanese cypress reuses the thing which used in a store before the remodeling. The design of the façade expressed an area in the restaurant. The logo of the signboard appears when the lighting lights up.

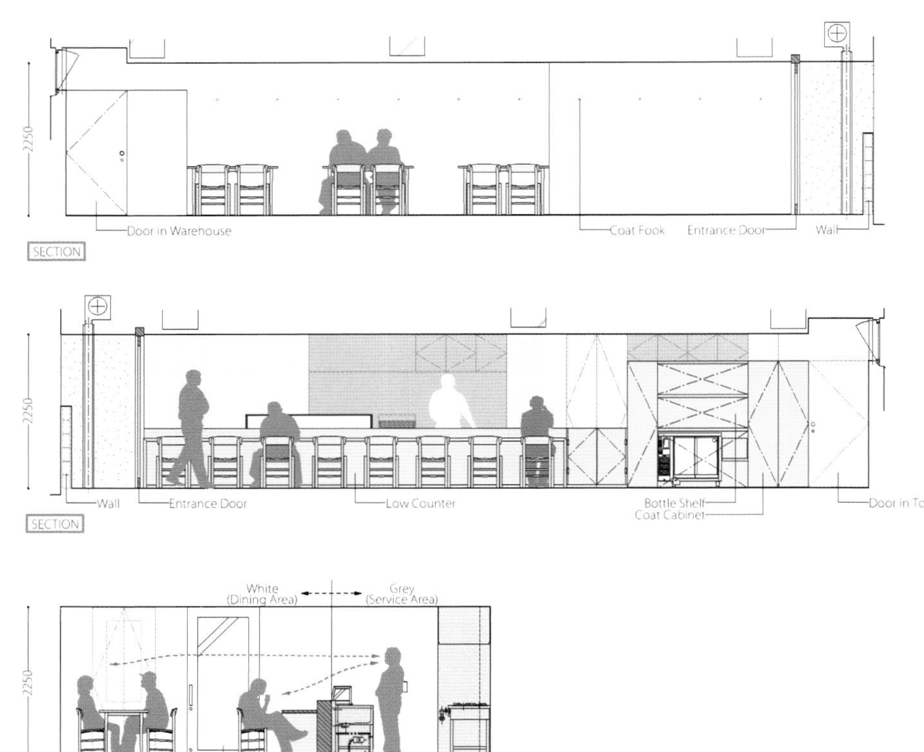
2250
Door in Warehouse
Coat Fook
Entrance Door
Wall
SECTION
2250
Wall
Entrance Door
Low Counter
Bottle Shelf
Coat Cabinet
SECTION
White
(Dining Area)
Grey
(Service Area)
2250
Entrance Door
SECTION

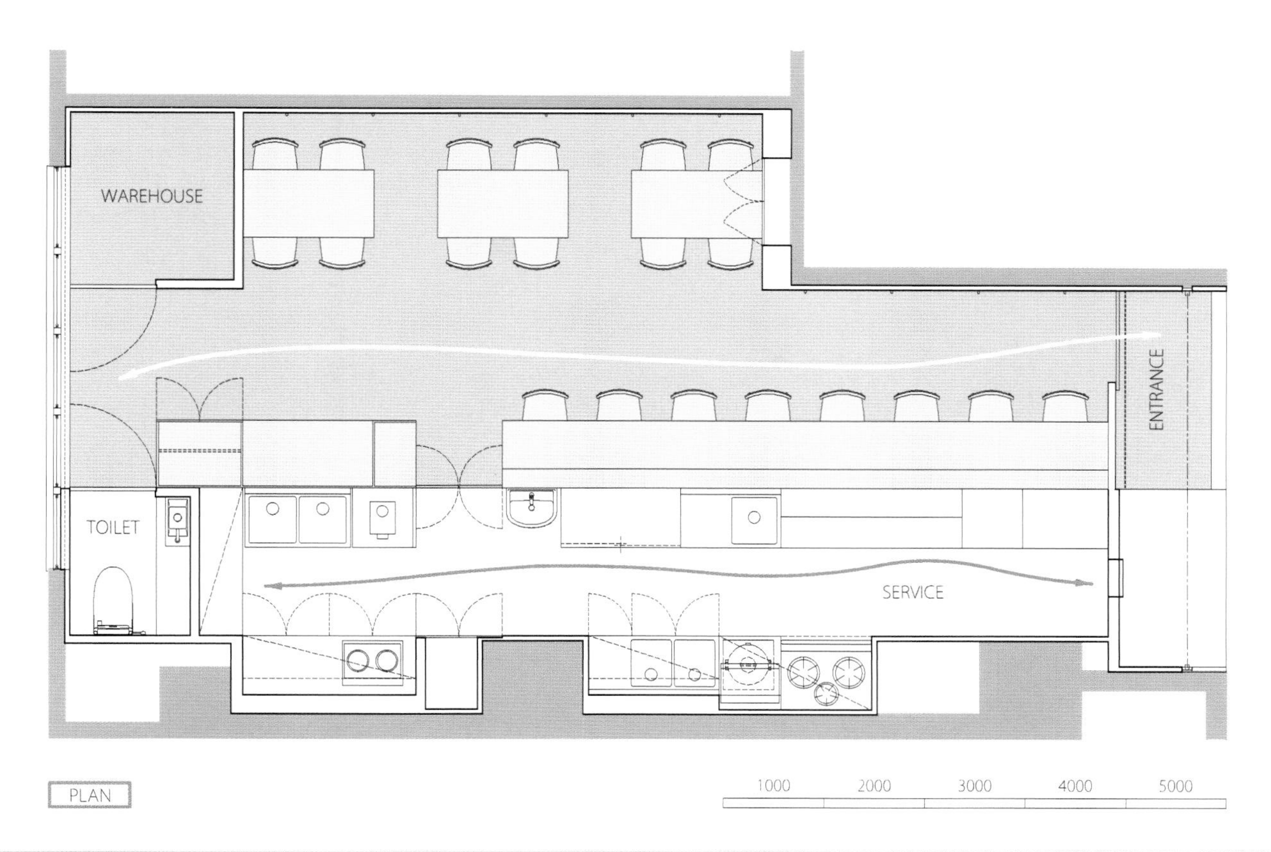
WAREHOUSE
ENTRANCE
TOILET
SERVICE
PLAN
1000
2000
3000
4000
5000

这个项目是翻新一家日本寿司店。

该店的一楼重新装修并于2010年10月开业。

设计师希望通过简单的室内装修和明亮的空间，能够让厨师更加高效地完成工作。同时，还要凸显寿司简单、干净的印象。

厨房的天花板和墙面都是重新布色，干净素雅。“简洁和突出整体”的设计令原有较低的天花板顿失压抑感。

柜台选材是日本柏树，是一家商店旧料的再利用。店面的设计传达着店内的信息。灯光亮起，店牌上的标志赫然在目。

PULPEIRA VILALÚA

Designer: Marcos Samaniego Reimúndez
Design Company: Mas • Arquiectura
Photographer: Ana Samaniego

The restaurant aims at bring guests the pleasure of good cooking.
Mas • Arquitectura moves to Madrid the essence of traditional galician "pulpeira". This is a unique place to share experiences and enjoy, at the same time, a tasty dinner. This is a "pulpeira" a traditional Galician restaurant where octopus is the main product. The project, developed for Marcos Samaniego from Mas Arquitectura study, adapts this traditional store to an urban space: Madrid.
The local, called "Pulpeira Vilalúa", follow a main idea: a place to share your time with family and friends. For this reason, the local has designed with continuous benches –an invitation to share experiences and perimetral layout. The furniture, designed by Mas Arquitectura team, bets for renewable materials, as old wood.
The entrance to "Vilalúa" universe supposes a trip towards unknown world: the pleasure of good cooking. A tunnel, which allows us to entry in the local, has been equipped with shelves and wastebaskets to turn this space into a comfortable place for smokers.
Inside, the local hides a lot of secrets. A special main door or wine barrels used as tables are interested details that should be discover.
Mas • Arquitectura, a Spanish young study, has developed this project from design to execution. Only this way of working guarantees a perfect harmony: furniture, lighting and layout are part of a set-"Pulpeira Vilalúa"

该项目的宗旨是让客人享受美食带来的快乐。

Mas•Arquitectura迁至马德里，是一个交流和品尝美食的好去处。它是一个传统的加里斯雅文餐厅，章鱼是其主要食材。应Marcos Samaniego的要求，对Mas•Arquitectura进行一番研究之后，汲取了传统商店的特点，并将其应用于马德里市区空间设计之中。

在设计Pulpeira Vilalúa酒店时，设计师始终遵循一个理念，那就是设计出一个可以跟家人和朋友分享美好时光的去处。为此，酒店的设计中便有了首尾相接的长凳，这为无障碍交流提供了积极的暗示。Mas Arquitectura团队设计的家具大胆地使用了可回收利用的建材，例如，旧木料。

Vilalúa“宇宙”的入口似乎是通向了一个神秘的未知世界：美食带来的无限快乐。一个装饰着草棚和废弃篮子的通道引领客人进入酒店，而且这里还是烟民们的乐园。

酒店里神秘之处也处处得见。一扇特别的大门、被用做餐桌的白色大桶，妙趣横生，让人应接不暇。

Mas•Arquitectura，一支年轻的西班牙团队，已经把这个设计作为公司执行任务的标准。只有这样的工作作风才能保证设计的完美、和谐——家具、灯光和样板房交相辉映，共成一体。

COCINA

HAIKU SUSHI

Designer: Edmond Tse
Design Company: Imagine Native Ltd. (Hong Kong)
Location: International Finance Center, Shanghai, China
Area: 320 m^2
Photographer: Kingkay Architectural Photography

The new Haiku Sushi is located at the open courtyard of the recently completed Shanghai International Finance Center. The designers utilize the concept of origami as the major driven force throughout the design. The restaurant space is sub divided in different zones, sushi bar, drinking bar, two main dining areas, booth seating and tatami rooms. Each zone is formed by an origami feature, which is constructed by colors and different materials, such as perforated aluminum composite panels, translucent stone panels and linen fabrics. With different lighting effects, each zone will have its own spatial character. These different zones in the restaurant are also tied by an origami suspended ceiling from the entrance to the VIP-rooms at the end of the shop. It unifies these feature zones and creates a spatial transition throughout the restaurant.

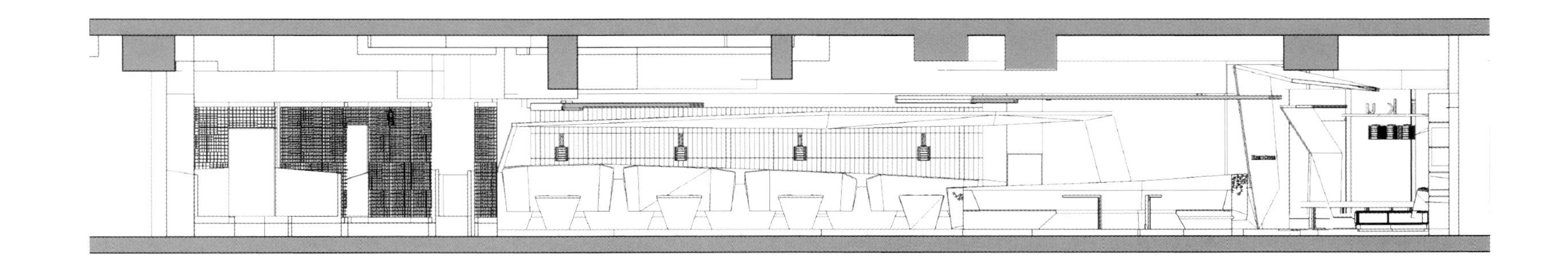

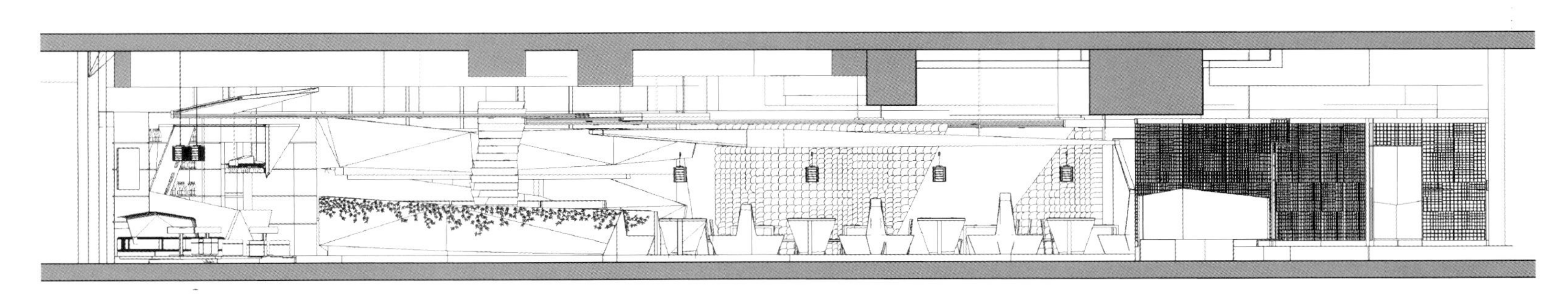

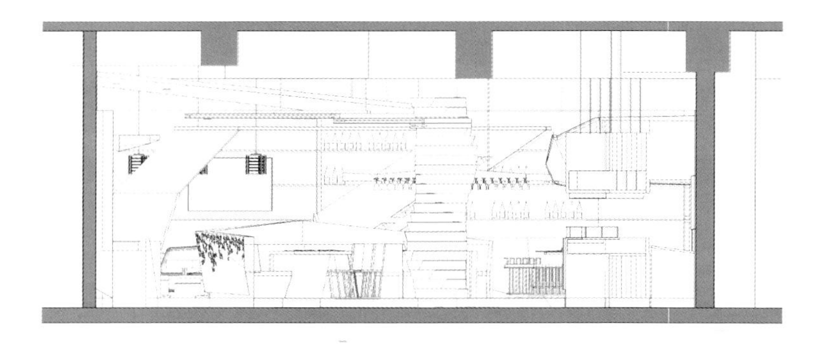

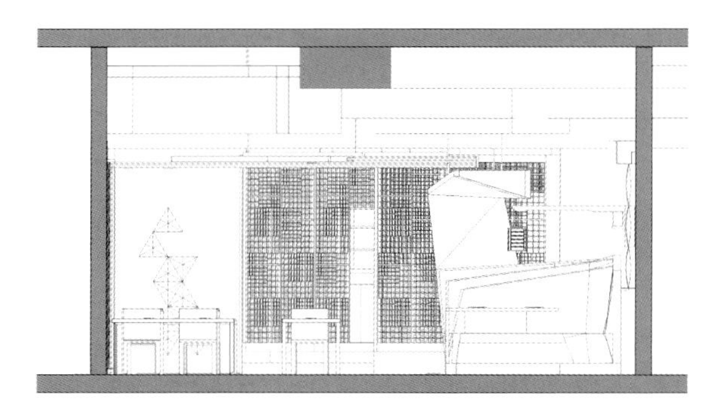

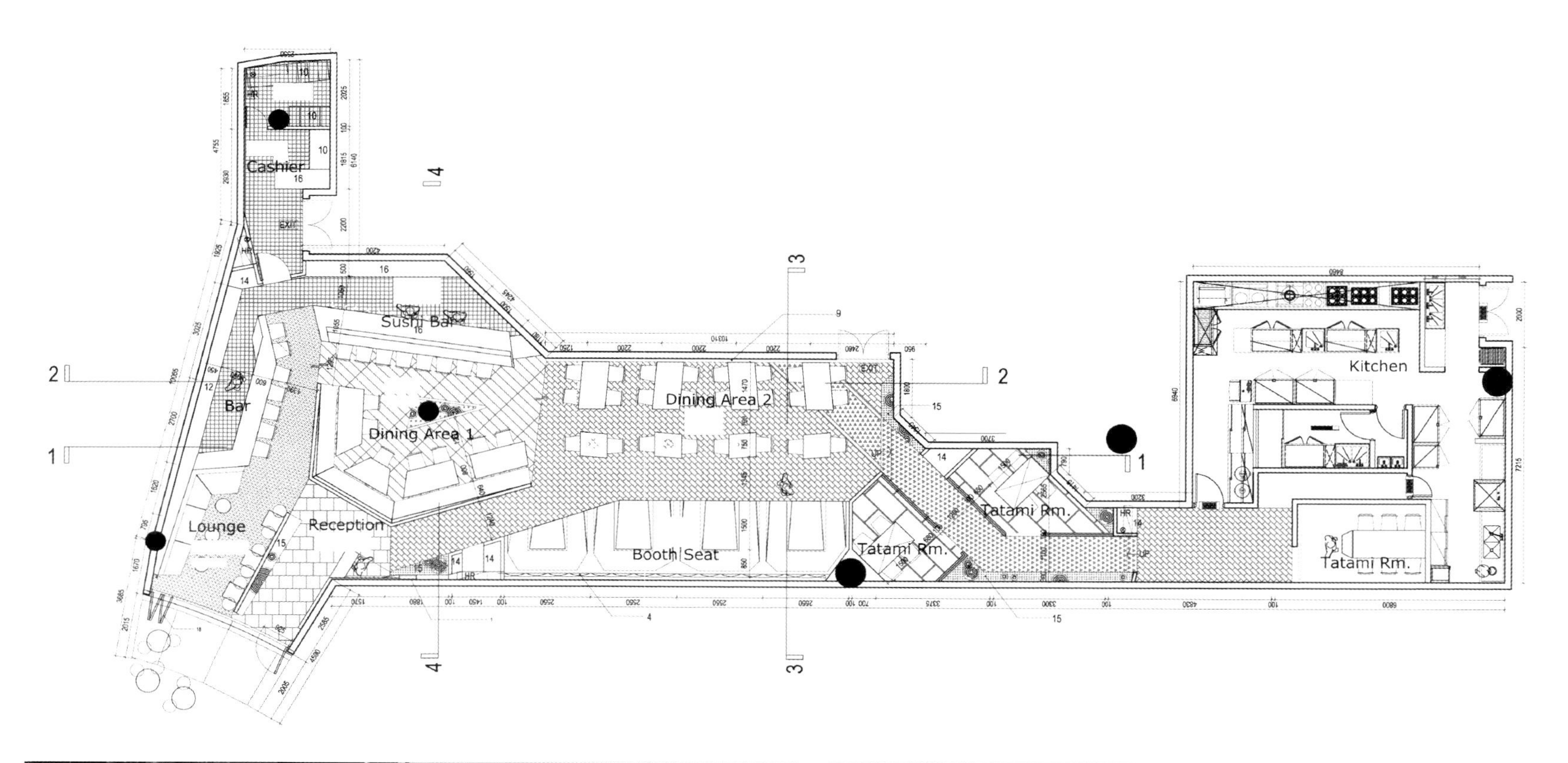
Cashier
EXIT
Sushi Bar
Bar
Dining Area 1
Dining Area 2
Lounge
Reception
Booth Seat
Tatami Rm.
Tatami Rm.
Tatami Rm.
Kitchen

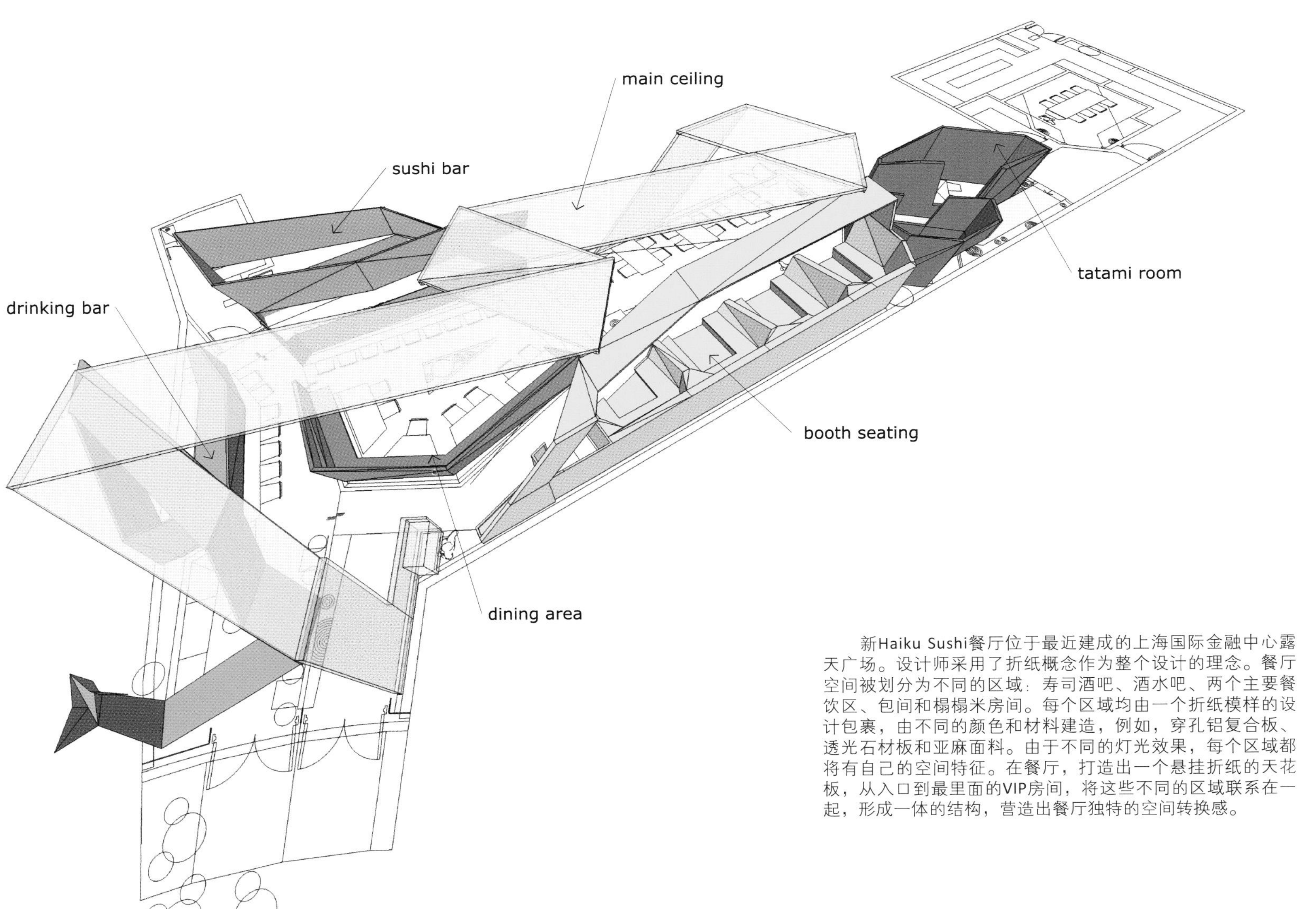

新Haiku Sushi餐厅位于最近建成的上海国际金融中心露天广场。设计师采用了折纸概念作为整个设计的理念。餐厅空间被划分为不同的区域：寿司酒吧、酒水吧、两个主要餐饮区、包间和榻榻米房间。每个区域均由一个折纸模样的设计包裹，由不同的颜色和材料建造，例如，穿孔铝复合板、透光石材板和亚麻面料。由于不同的灯光效果，每个区域都将有自己的空间特征。在餐厅，打造出一个悬挂折纸的天花板，从入口到最里面的VIP房间，将这些不同的区域联系在一起，形成一体的结构，营造出餐厅独特的空间转换感。

YABANY

Designer: Brunete Fraccaroli
Location: São Paulo – Vila Olímpia, Brazil
Area: 150 m²

The architect Brunete Fraccaroli creates, in this project, an unusually japanese restaurant, she abuses from the freedom of the materials, mixing steel and glass, such as Wood and natural fibers.
The concept begins mainly with the strong use of color, predominating the orange and the pink. The attractive game between these colors, added to the transparency of the glass and acrylic create a large space, contemporaneous and "clean", without escape from the Japanese tradition. One of the most preoccupations of the client was not to loose the oriental concept, but the architect solves that in a single and original way.
The old property was demolished and replaced by a hangar with PÉ DIREITO DUPLO(double height). The project contains a mezzanine with natural light that involves harmonic and comfortably the place.
In the choose of the furnishing, the architect shows one of your best characteristics, the use of technology, added to the colors, which, at the same time, refers either to the illumination project and creates a nice place, audacious, innovative, with unexpected angle, original forms, amazing transparencies, without loosing the reference and concept of the project, which is a place related to the Japanese culture.
In this Project, the architect can elaborated an audacious room, where the rich details make the difference. The colorful luminaries composes the place and increases the chromatic idea, principally in the colors orange and pink.
The colors and the light are the strong points of the project and the architect, in a brighter way explored this concept with the uses of glass, which won´t give the same effect to the set, if were used only transparent.

在该项目中，设计师Brunete Fraccaroli建造了一家非同寻常的日式餐厅。她灵活运用各种材料，将钢材和玻璃进行组合，例如木材和天然纤维。

设计理念开始于强烈色彩的应用，主要是橙色和粉色。这些颜色的有效搭配，突出了玻璃和亚克力材料的透明度，营造了开阔、现代、整洁的室内空间，使其具有日本传统和现代相结合的特征。客人最关注的就是餐厅不要失去东方的特色，而设计师则以简单且原始的方式解决了这个问题。

原有的房子被推倒，被一个具有PÉ DIREITO DUPLO的飞机库所替代。这个项目还包括一个自然光可以照射到的夹楼，营造出和谐、舒服的空间氛围。

设计师在装饰餐厅时，通过技术的运用和色彩的添加，打造了一个漂亮、创新的空间。在保留建筑理念的前提下，运用了出乎意料的角度、原始的形状和令人惊叹的透明度，建成了这个与日本文化息息相关的餐厅。

在该项目中，设计师打造出一个创新的空间，丰富的细节使其别具一格。多彩的照明灯具点亮了整个空间，增加了人们对彩色的体验，尤其是橙色和粉色。

色彩和灯饰是该项目最大的亮点，并且建筑师通过玻璃的应用，以一种更加通透的方式实践了这个理念。如果单纯使用玻璃的透明度，就实现不了现在的背景效果。

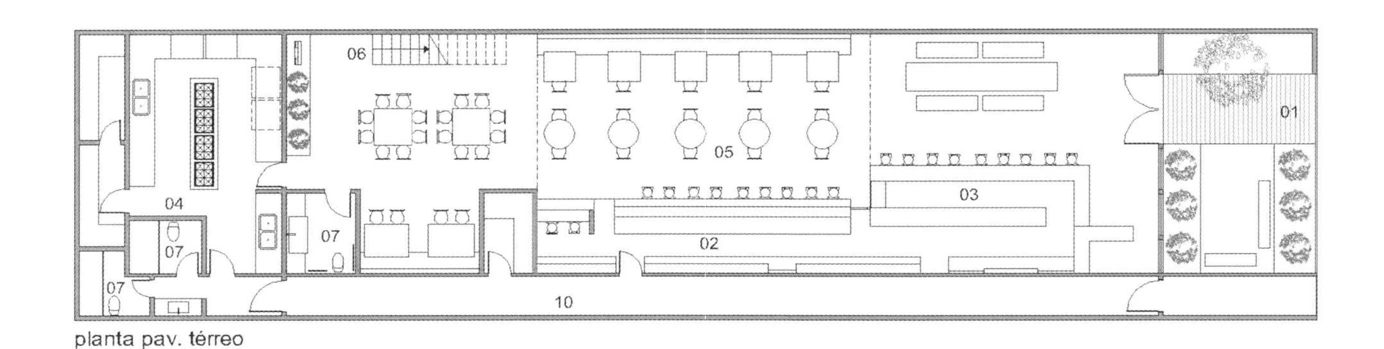

planta pav. térreo

planta mezanino

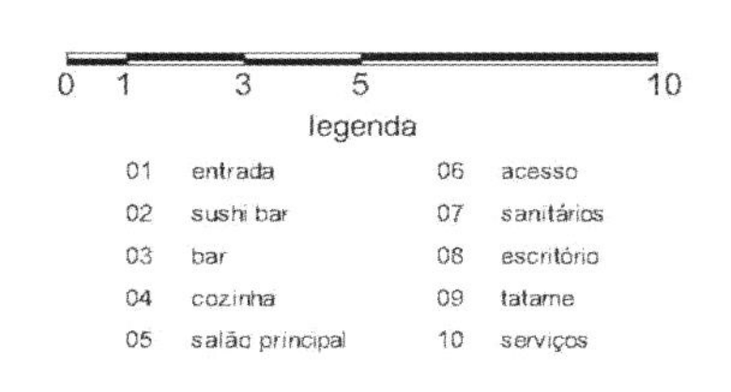

KAREN TEPPANYAKI, TAIMALL

Designer: Frankie Yu
Design Company: Gure-Design
Area: 132 m^2

The designer made good use of strong black and red to create a brand-new image, combining robust with delicacy, conveying the extreme contrast of tradition and modernity.

Lines on the red glass wall with surface finish of paining in the center of the restaurant are extremely clear. The other wall is covered with black and white wallpaper, expressing a variety of architecture languages. Tiles of charcoal and black covered the floor collage, matching the dark colors of red, black and silver gray. There is a huge column between the two grills. The designer covered it with gray collage mirrors, the spaces reflected from the mirrors weakening the column itself and enlarging the space.

For dinners, watching the skill of main chef is quite important when enjoying delicious meal. So it's necessary to add spotlights above the chef table to the pendent lamp, which enables the dinners to appreciate the consummate skill of the chef. Arc grill facade lit by red LED light, strong black and red are developing unique decadent sentiments.

紧急出口

Asahi

EXIT
紧急出口

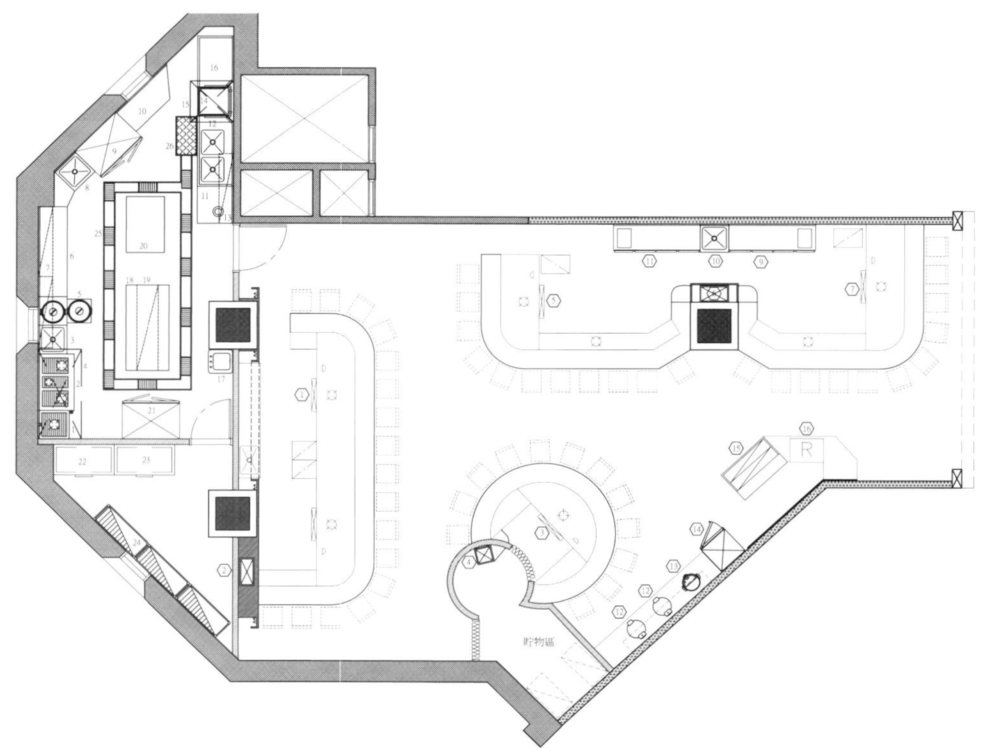
貯物區

设计师打破传统刻板制式的印象，大胆运用黑、红两种强烈的色彩，糅合粗犷和细腻，展现出现代古典极致的冲突美感。

餐厅的正中央主墙面以红色烤漆玻璃分割磨边的方式，呈现出线条层次感。另一面墙面则以黑、白色的花纹壁纸铺陈，凸显出丰富的空间表情。地面选用铁灰色和黑色两种颜色的进口板岩砖拼贴，呼应整体的红、黑、银灰等暗色调；具流线感的弧形天花区隔铁板烧区，铁灰色系与地坪相互呼应。两个铁板烧台中间有个大柱子，设计师以灰镜拼花贴覆，镜子反射出其他空间，相对虚化了本身的柱体，同时也放大了空间。

观看厨师制作铁板烧料理的功夫，是进餐的过程中重要的一环，因此在灯光的配置上，除了吊灯的重点式照明外，在主厨料理的位置上还加设投射灯，每位客人在用餐的同时，都能欣赏一场由厨师带来的精彩美食秀。圆弧的铁板烧台面透出红色的LED光线，黑和红色的大胆的结合，诠释强烈又有个性的颓废风。

OCEAN ROOM

Photo I. Susa

Designer: Yasumichi Morita
Design Company: Glamorous Co., Ltd.
Location: Circular Quay W, the Rocks NSW 2000, Australia
Area: 536.2 m^2
Photographer: Sharrin Rees

Ocean Room is a restaurant located in Sydney Cove with a beautiful view of the Opera House.
The designers were asked to create an atmosphere that was suitable for a Japanese modern restaurant, so they chose a wood material as the defining motif for the space. The ceiling space has been dramatically reborn with 42,458 beads of warm Asiatic-style wood hanging in beautiful dynamic ways toward the floor. As the air rustles through the wooden wave formations, a relaxing atmosphere of gentle movement and weightlessness is experienced, a new combination of Asian wind and wave.
The atmosphere is further enhanced with the lighting plan delicately designed so that each dish and guest is beautifully lit.
The layout has also been revised to create a stylish new bar area. Svelte black tiles deck out the bar, while lavish warm wooden floorboards facilitate the dining area. This was designed to develop and separate a unique bar or dining experience that guests may enjoy in either area. Moreover, the beautiful design of the bar can be seen from outside the restaurant through a floor-to-ceiling glass wall, allowing the atmosphere of the bar to allure and entice people to enjoy what is on display.

Ocean Room是位于悉尼湾（Sydney Cove）的一家餐厅，在此能看到悉尼歌剧院的美丽景色。

创建餐厅时，业主要求打造出日式现代餐厅的氛围，所以，设计师选择木质材料来装饰整个空间。天花板上悬挂着42 458颗小珍珠串联成的亚洲式样的木条，以漂亮、动态的方式垂向地板。当空气流动时，就形成了木质的波浪，营造出轻柔、移动、放松的气氛，让人体验到失重的感觉，这是亚洲风和浪的新的结合方式。

精心设计的灯光照亮了每一道菜品和每一位客人，更为餐厅注入了活泼、奇特的气息。

为了打造一个时髦的新吧台区域，餐厅布局也曾做过修改。清晰的黑色瓷砖用来装饰吧台，奢华、温暖的木质地板铺设在就餐区域。这样设计的目的是将客人在酒吧或者就餐区所享受到的体验区分开来。而且，酒吧漂亮的设计从餐厅外面的落地玻璃窗就能看到，这样可以吸引客人进来体验在此进餐的美妙。

APOSTROPHE

Designers: Neil Hogan, Brendan Heath, Adam Woodward
Design Company: SHH
Location: London, UK
Area: 155 m^2
Photographer: Francesca Yorke

The site was challenging as it was long and narrow. The solution for the site involved a careful division between operational and customer areas. Several seating zones were created, each with a unique feel and furnishing arrangement. In addition, the space was broken up into operational areas for both 'grab and go' and assisted service, as well as the wash-up zone, food prep and bakery area.

Fascia and projected signage were both needed to give the narrow site as much visibility as possible. Glazing is full-height so that the first 'lounge' seating area to the right of the entrance becomes a living advert for the brand and interior. At the left side of the glazing, the first "assisted service" display case is prominent and highly visible to passers-by.

Flooring in the lounge area and throughout is a grey ceramic. Furniture in this area is freestanding. Also visible from the exterior in the lounge area is the "Clouds" decorative wall tiles, configured to run up the wall and to wrap over the ceiling. The wall hanging system is in a colourway, which fitted perfectly into the scheme. It also serves as a good sound-absorber because of the faceted surfaces of the individual components. The chilled areas are made out of stainless steel with UV-bonded glass in order to minimise visual structure and maximise the display of products.

The walls throughout are either in American black walnut panelling or black laminate with a high datum level and dark grey paint up to the ceiling, where services are left exposed and original remains of plaster cornices are also simply painted grey.

The central zone houses further banquette seating, upholstered in black, as well as more freestanding seating, with feature lighting above. Deeper into the unit , in front of the bakery area, is a large, freestanding solid walnut meeting table, which can seat up to eight people and is there to encourage 'breakout' meetings from local offices, as well as to break up the zones as a feature area. Behind the table, this time against a bright white wall panel, is the second use of the 'Cloud' wall tiles, in a slightly smaller formation. The area is also bordered by a set of newspaper slots lined in gloss pink Chameleon from Altro and set into the walnut wall at one end. At the far end of the wall is one of the scheme's major design features: the pink glass bakery area.

apostrophe
the accent on taste
apostrophe
winter
soup
range

apostrophe
apostrophe
apostrophe
espresso · macchiato
cappuccino · latte macchiato
latte · filter coffee · americano
hot chocolate · mochaccino
chai latte · cinnamon latte
loose leaf teas & herbal blends

apostrophe

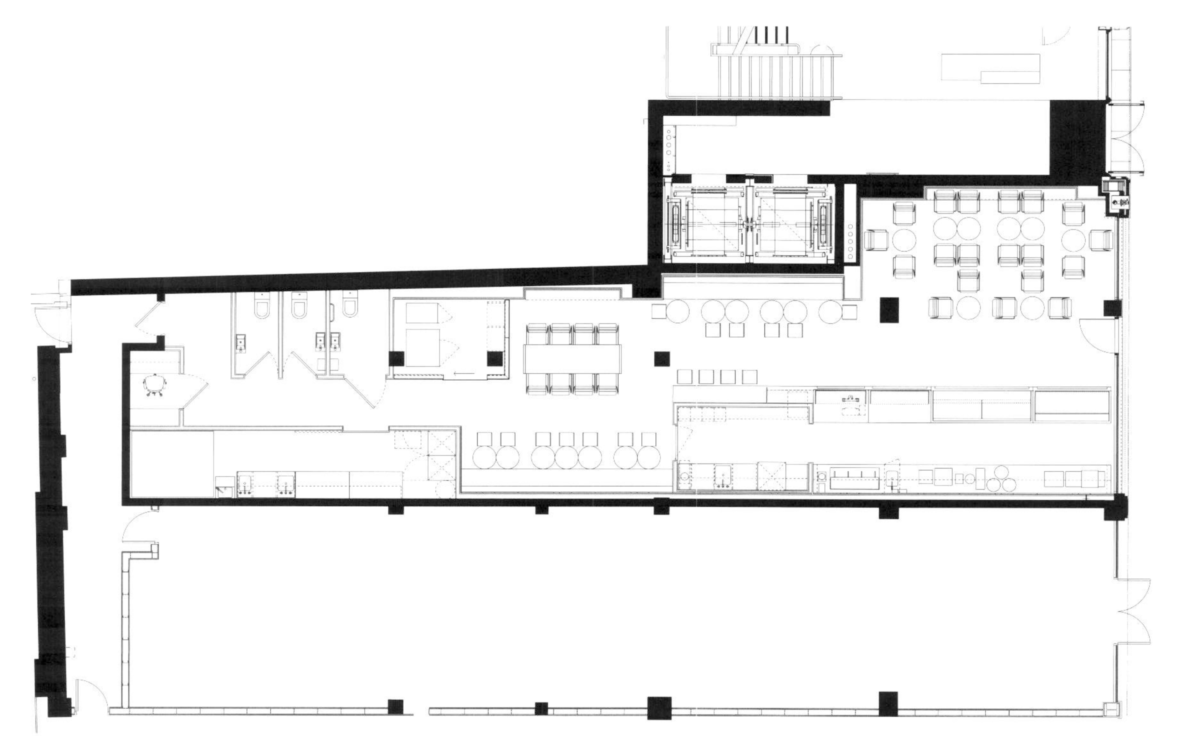

原空间的狭长布局是此次设计的挑战，所以设计师把操作区和客人区进行了细致划分，打造出多个座位区域，通过不同的陈设营造不同的氛围。并且，这些区域有效划分为：外卖和辅助服务区、洗涤区、备餐区以及烘焙区。

招牌设计也给这个原本狭窄的区域带来足够的吸引力。店面主体由全高的镶嵌玻璃构成，客人可透过玻璃窗一眼看到入口右边的休闲座位以及左边提供“辅助服务”的展示架，达到自我宣传的效果。

灰色的瓷砖为室内地板的主材料，家具都呈独立式摆放。从窗外可看到休闲区醒目的云状墙饰，它如漂浮的彩云沿墙飘升直至天花板。这款云状墙饰不但以彩色的搭配而实现了与整体空间的完美搭配，同时其块状的表面还能带来吸音的效果。

左侧主柜台的外面是洗涤区域，是一个独立的盒状空间，颜色采用的是招牌的颜色，外部是黑色的薄板，内部是平滑的粉色薄板。

中央区域放置的是供客人休息的黑色带垫椅子，还有更多独立式的座位，上方悬挂着特色的照明灯。在餐厅的深处、烘焙区的前方，有一个大型的独立式胡桃木会议桌，这张会议桌可坐8人，可供周边的办公人员作为临时会议和休闲的场所。会议桌后面的墙壁则是以白色作为幕布，幕布的中央再一次运用了漂浮的“彩云”，形成了一幅完整的框画作品。会议区一旁的胡桃木墙上嵌有一排粉红色框架的报纸插槽。而以粉红色玻璃幕墙为特色的烘焙区则设在这面墙的远端。

PAPAPICO

Design Company: Atelier Heiss Architeckten, Vienna
Area: 180 m^2
Photographer: Peter Burgstaller

Hanging lights made of felt hover in the air like warming food covers. A vine pattern on the ceiling lets guests make a fictive escape to a Mediterranean vineyard. These are only two of the highlights in the design of the new "Papapico" cafés by Atelier Heiss. The building and the design concept by Atelier Heiss captivate with clear orientation and a strong outward identity. On the inside, however, the chosen materials are warm but simple. The flagship store in Vienna -Floridsdorf will be the new appearance for all coming cafés. Contrasts range from peace and quietness to liveliness. The pale green accents on the chairs and the extensive use of wood give a feeling of relaxation and joie-de-vivre. The eye-catching bar with its LED-lighting becomes the centerpiece of the room.

coffee

pane
pasta
pizza
coffee

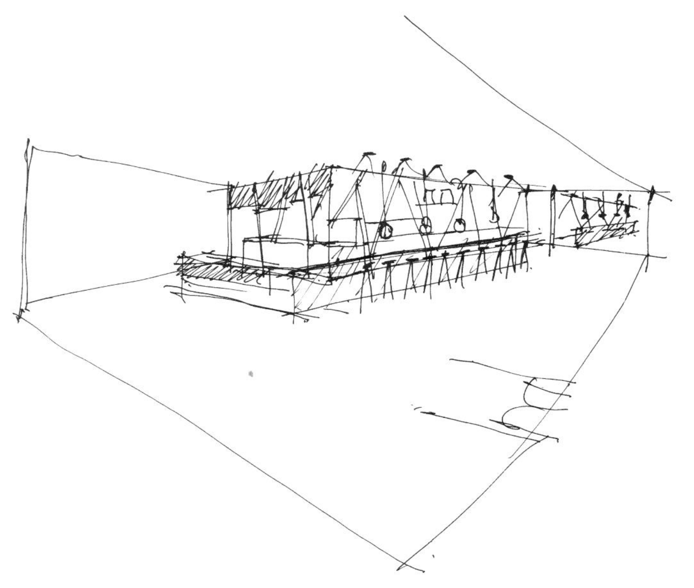

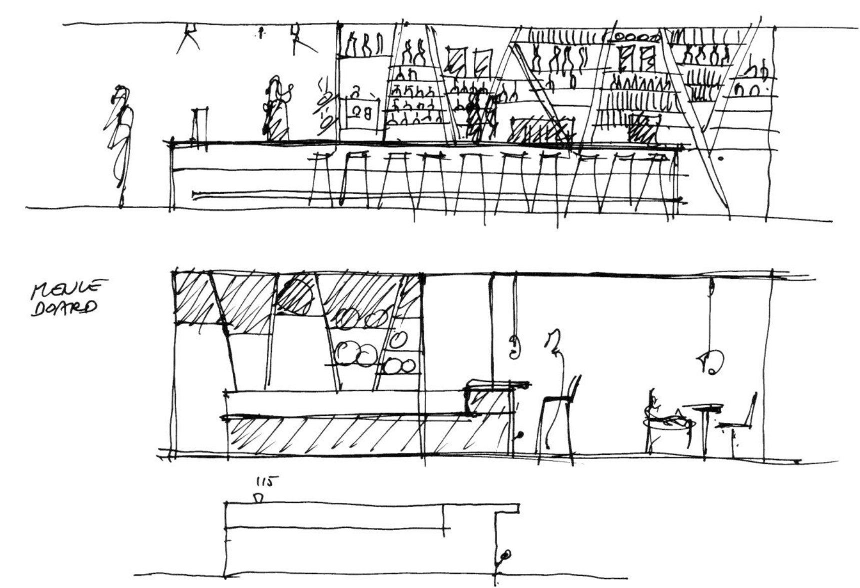
MENUE BOARD
115

餐厅中的吊灯悬挂在空中，就像是为食物保暖的盖子。天花板上的蔓藤花纹让客人有一种置身地中海葡萄园的幻觉。以上是Atelier Heiss设计公司在设计Papapico咖啡馆时的两个亮点。

Atelier Heiss的建筑设计理念是方向明确以及个性鲜明。在内部装饰方面，所选的材料既暖和又简洁。在维也纳的Floridsdorf旗舰店内，陆续会有许多咖啡馆全新亮相。

具有反差的设计风格从恬静到活泼。餐厅内部淡绿色的木制椅子，给人一种放松和“偷得浮生半日闲”的感觉。引人注目的LED背景光吧台，成为餐厅的焦点。

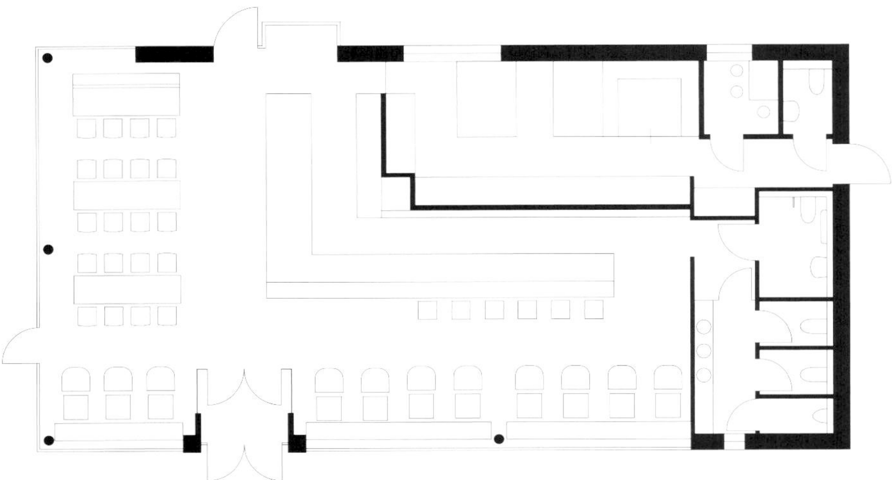

TIPICA RESTAURANT

Design Company: Klm-architects
Location: Berlin, Germany
Area: 130 m^2
Photographer: Klm-architects

In February 2010, a new gourmet-restaurant opened in Berlin, the capital of Germany. Klm architects designed this Tipica in a modern mexican style with elegant white walls, purple glass, pink accessoires and warm natural colours.

Center of the restaurant is a shiny-golden bar with a jointless surface of white mineral material. Parallel to this bar a high glass wall accompanies the visible kitchen. The white walls, ceiling and leather upholstry glow like the typical mexican houses besides sandy-colored surfaces. Six big hanging lamps reflect golden light-the lucent lamp shades above each table remember of corals. Glass-cubes with the typical flavours and spicy chillis are presented on a silky wallpaper and complete the culinary experience.

The team of klm-architects - Olaf Koeppen, Sebastian Leder and Rita Märker – combine the mexican vitality with the elegance of Berlin. They developed in a close collaboration with the gastronome an extraordinary restaurant, which converts the ease of Latin-America into an efficient gastronomical and architectural concept.

Bright and inviting in the day-time while elegant and mysterious in the evening, a strong corporate-design for other Tipicas will continue in the future.

2010年2月，一家新的美食餐厅在德国首都柏林开业。Klm公司的设计师以一种现代的墨西哥风格设计Tipica餐厅。他们大胆运用了优雅的白色墙壁、紫色玻璃、粉红色装饰和亲近自然的色彩。餐厅中心位置是一个将无缝的白色矿物材料作为表面的金色吧台。

吧台的旁边是一个高的玻璃墙，里面是厨房。除了沙色的表面之外，白色的墙壁、独特的天花板设计、这一切都是典型的墨西哥建筑的装饰风格。6个大挂灯反射着金光。每张桌子上方的朗讯灯罩，都会让人联想到珊瑚。装有经典调料和辣椒的玻璃立方柜，呈现在柔滑的墙纸上，让客人的美食体验更加完美。

Klm公司的设计师Olaf Koeppen、Sebastian Leder 和 Rita Märker通力合作，将柏林的优雅墨西哥的活力完美结合在一起。他们通过将拉丁美洲的美食风格和建筑理念巧妙融合，从而打造了一个美食家心中的非凡餐厅。

餐厅日间宽敞明亮、热情好客，晚间则略显优雅神秘。Klm公司将在未来设计建造更多其他的Tipica连锁餐厅。

NISEKO LOOK OUT CAFÉ

Designer: Yuhkichi Kawai
Design Company: Design Spirits Co. , Ltd.
Location: Hokkaido, Japan
Area: 172 m^2
Lighting Consultants: Muse-D Inc. Kazuhiko Suzuki, Misuzu Yagi
Photographer: Toshihide Kajiwara

The Niseko Look Out Café features rows of lattice-work booths and slatted timber false ceilings. Partitions were also made from timber lattice. Only three types of materials including woods, paint, and wallpaper are used for designing the restaurant. Vertical timber lattice has been used as the main material to lend an exotic feel to the space with reflections coming from the roofs come in different sizes and heights. Illumination effect is created from the lights from roof through the lattice while the space is surrounded by roofs with various sizes and heights. The feeling is more obvious when the space is crowded.

Niseko Look Out咖啡厅的特点是一排排格架式亭子和板条式木质装饰天花板。各个分区也是木质格架结构。本咖啡馆的装饰选材仅限于三种：实木、油漆和墙纸。

将竖直的实木格板作为主要建材，自天花板而下，宽窄不一，高低错落，呈现出一股奇特的异域风情。

天花板上的灯光穿过宽窄不一、高低错落的隔板，显得更加明亮，这种效果在客人较多时尤其明显。

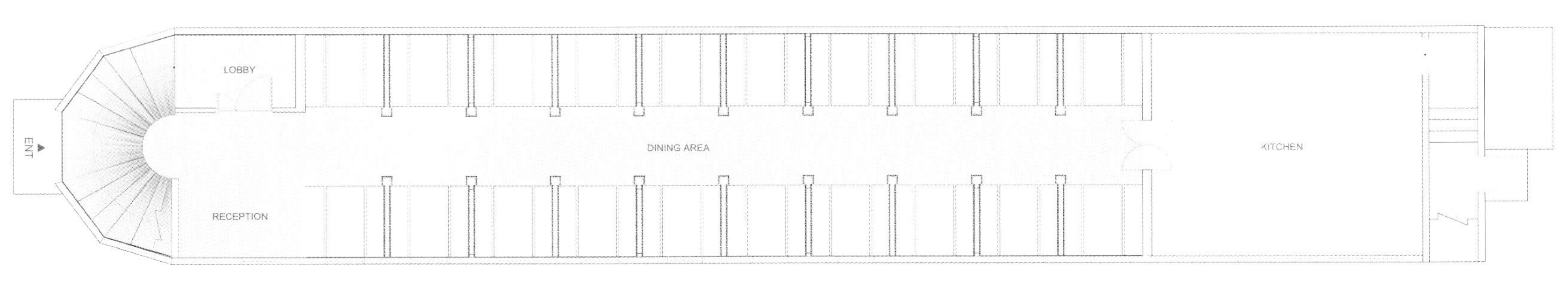
LOBBY
ENT
DINING AREA
KITCHEN
RECEPTION

OKKU

Designer: LW Design Group
Design Company: LW Design Group
Location: Monarch Hotel, Dubai, the United Arab Emirates
Materials: Dark Stained Timber, Black Wrought Iron, Honed Granite

Okku, is the latest addition to the achingly trendy night scene in Dubai. This Japanese restaurant / nightclub in the Monarchy Hotel opened its doors to Dubai's glitterati in 2009 and it has fast become the place to see and be seen. Taking centre stage above the bar is a mesmerizing aquarium filled with pulsating jellyfish, adding a hint of atmospheric danger to the dark and sultry interior scheme. Water features reflect the flicker of hanging candles and lit curtain dividers provide a moody light and a degree of exclusivity to the edgy space.

The contemporary Japanese design is created by using dark, robust materials and retaining the sense of mysticism and exoticism evident in modern Japanese interiors. Black wrought iron screens and balustrades compliment the dark stained timber and honed granite panels, while the circular white leather seating introduces a more playful vibe to the distinctive design.

JIM BEAM
竹鶴
白州

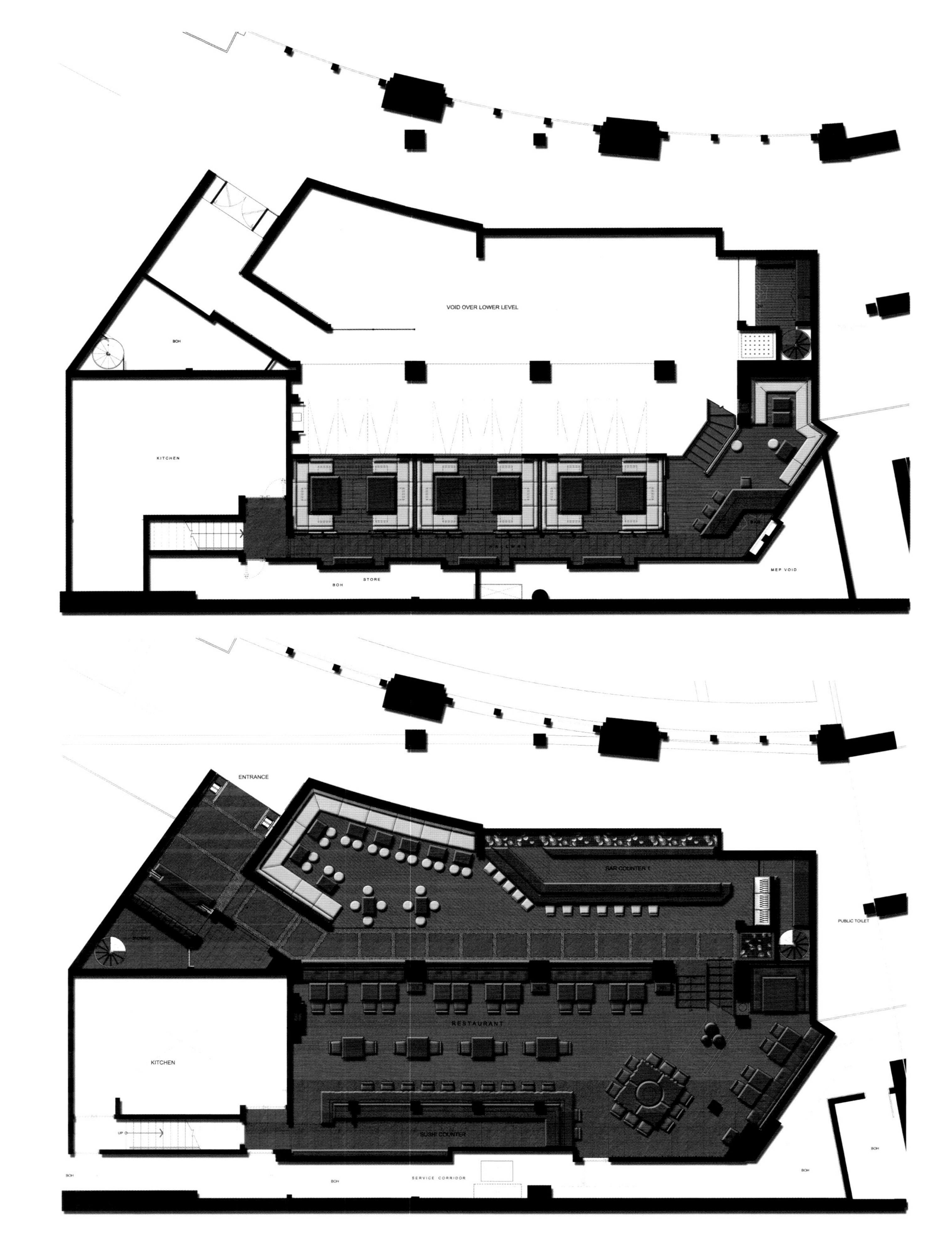
VOID OVER LOWER LEVEL
BOH
KITCHEN
BAR
BOH
STORE
MEP VOID
ENTRANCE
BAR COUNTER 1
PUBLIC TOILET
RESTAURANT
KITCHEN
UP
SUSHI COUNTER
BOH
BOH
SERVICE CORRIDOR
BOH
BOH

在闷热躁动的夜晚，置身于引领潮流的烛光音乐会，摆着时髦的姿势，品着日式鸡尾酒，享受着Okku——在迪拜，这又是一种引领时尚的夜生活。Okku是迪拜时髦夜生活的一个有力补充。这家日式餐厅/夜总会位于迪拜君主酒店内，在2009年面向社会名流开业，很快就成为既养眼又露脸的好去处。吧台上面的中央位置放置着一个迷人的养鱼池，里面养着游动的水母，提醒人们黑暗、性感的内部环境中暗藏的危险。水能反射出悬挂蜡烛的摆动，窗帘的起落能够为客人提供变幻的灯光和对私密空间的独享。

当代日式设计采用的是深色的、坚固耐用的材料，并保留了现代日式内部设计的神秘感和异国情调。黑色的、加工精致的铁影壁和栏杆与深色的木材和花岗石嵌板相得益彰，而圆形的白色皮质座椅却为这个独特的设计营造出了更加有趣的氛围。

SUBU

Designer: Johannes Torpe
Design Company: Johannes Torpe Studios
Location: Beijing, China
Year: 2008

SUBU is made by the family behind "South Beauty Group". Owning more than 50 restaurants and lounges all over China, they are one of the largest and most influential restaurant operators in China. South Beauty Group wanted the designers to rebrand, redefine and redesign the "classic" Chinese restaurant. This should be done in a completely new way yet it should still maintain its roots and references to Chinese culture and traditions.

Vision

The designers wanted to make something exclusive, spectacular and outstanding by designing all aspects of the experience from the definitions of the space to the chairs and the cutlery. The grandness of the budget and the scale of the project allowed to carry the vision through, creating a spectacular dining experience.

Idea and concept

The designers were very inspired by some of the traditional Chinese restaurant features such as the private rooms. They took the Chinese inspiration and fused it with the Scandinavian design approach and created an atmosphere that is international yet accentuated with Chinese influences.

What the designers wanted was to give the place an 'airport feel', making it visible from the whole mall – and at the same time create an intimate and private dining atmosphere. They did this by creating 5 arcs that fold itself protectively over the dining area. The arcs function as a significant visual statement in the grandness of the space bringing the roof down enclosing and defining the restaurant area.

For dining in an exclusive and intimate atmosphere, the designers developed the Cocoons. They are this century's version of the classic Chinese private room. Enclosed, shiny, sealed and generic on the outside while invitingly comfortable and unique on the inside.

SUBU是“俏江南餐饮有限公司”旗下的一家连锁店。在中国，俏江南拥有五十家餐饮休闲店面，是中国最大、最具影响力的餐饮机构。俏江南希望通过设计师重塑品牌、重新定位并重新设计“古中国风”的餐厅。这就要求设计师不但要以一种全新的方式来重新定位，而且还需保有其根植于中国文化传统的风格。

视觉效果

从对空间的定义到座椅再到餐具，设计师试图在各个方面展现俏江南的富丽堂皇和卓尔不群。充足的预算和工程的规模促使全方位的整体设计得以顺利进行，从而打造出一个华丽的餐饮场所。

思路和概念

传统的中式餐厅的鲜明特色激发了设计师的创作灵感，例如，包房。将中国特色和斯甘底那维亚的设计方法相融合，从而创造了一种国际化的氛围，但又突出了中国特色的就餐环境。

设计师试图展现一个私密的就餐环境——“身在机场”，视野开阔，同时不受外界的干扰。这个效果是通过架构在餐厅上方的五个圆拱相互支撑而完成的。圆拱自屋顶直垂地面，一气呵成，成就了餐厅的恢宏、大气。

为了营造一个尊享私密的就餐环境，设计师设计了茧状的包间，它们是传统的中式包间在现代社会中的呈现，外观神秘而又色彩炫目，貌似封闭却又活力无限；内部舒适却又独具匠心。

DELICATESSEN

Designer: nemaworkshop

For the design of Delicatessen, nemaworkshop explored the concept of urban identity, namely, the vibrant SoHo neighborhood and more specifically the NY newsstand. Typically overlooked because of its ubiquity, the newsstand serves as an atypical yet appropriate model. They are open, in-and-out thoroughfares which foster a unique breed of social energy. In order to capture this energy, the main dining space opens up to the bustling street corner with large retractable steel-and-glass garage doors. Traditional boundaries between street and restaurant evaporate, allowing the main dining space to be loosely defined and astonishingly inviting. Underneath the bar stands a collection of clear soda pop bottles while behind the bar, the wall is clad in leather subway tiles.

The concept of informal urbanity develops complexity as one travels downstairs into the more intimate spaces below street level. The existing boiler room has been transformed into a glass-roofed lounge lined with a mural by local artist Juan Jose Heredia. Just a few steps down from the lounge / courtyard is the minibar. As the name implies, it's a cozy underground bar where the walls are lined with vintage mini-bottles. Ultimately, the design of Delicatessen from the multi-leveled spatial organization to the comfortable yet sophisticated materiality is rooted in the concept that the space should be both inviting and accessible.

DELICATESSEN

RZA
MASTA KILLA

针对Delicatessen，设计师巧妙地借用了城市中特有的概念——充满活力的SOHO大街及其特有的纽约报亭。由于报亭普遍存在于城市的各个角落，所以设计师常常会忽视它们的存在形式，但它们却以一种非典型但又非常独特的形式融入社会。设计师看到了报亭外放，而且对街区特有的临时性等特点，这种社会交际场所，将忙碌在纽约的人们联系到一起，而这正是Delicatessen所需要的。外放的概念被置入主用餐区，面对熙熙攘攘的街道，设计师将废弃车库大门的回收钢化玻璃作为落地窗，表明Delicatessen的决心——融入SOHO区。在这里，街区与餐厅之间传统的分界线消失了，室内精美新潮的艺术透过玻璃窗展示给公众。餐厅的临街层以其高光材料与迷人的灯光突出室内艺术家的绘画艺术。类涂鸦的扭曲人面像以背板的形式铺设在吧台底层区域，绘画前成列摆放着透明的玻璃酒瓶，与吧台的各色酒品辉映成趣，显示出Delicatessen繁多的酒类与年轻个性的氛围。一层贴心地设置了台座与标准的四人用餐桌，给来此用餐或仅为消遣而小酌一杯的客人都提供了合理的活动区域。

经由一层的通道循序向下，通过SOHO楼体内的中庭区域就是地下的纯粹酒吧区了。步入地下部分，上层非正式的设计即笼罩上一种复杂的气氛。废弃的锅炉房改造成带有玻璃屋顶的休息区，配有一幅来自当地艺术家Juan Jose Heredia的超大型壁画。通过中庭的休息区仅几步之遥就来到了迷你吧。正如其名，这里是一个舒适的地下酒吧，墙体的设计格外突出酒吧特点——主体为线性排列的老式迷你葡萄酒瓶。从多层的舒适空间的组织构架到非正式高雅的艺术空间，设计师的创作思想还是根植于空间概念本身，为来此的客人提供一种既吸引眼球又不受束于艺术的舒适感。Delicatessen完全迎合了纽约本土的价值观。

STREET LEVEL

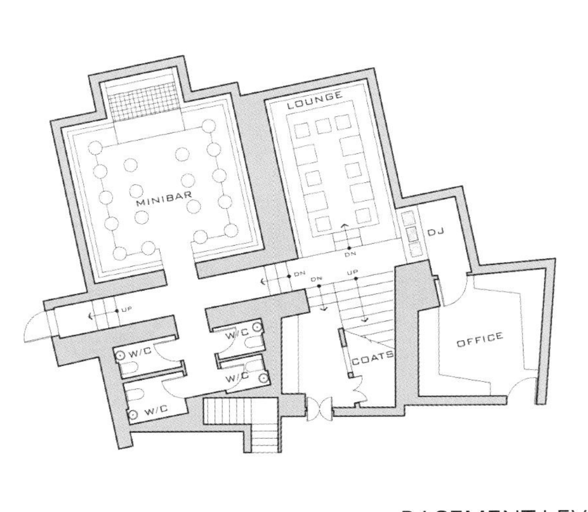

BASEMENT LEVEL

XIV

Designer: Philippe Starck

This October, one of the country's preeminent chefs, Michael Mina, will collaborate with SBE to debut the restaurant XIV, an innovative experience in dining that promises to add an exciting new dimension to the Los Angeles culinary landscape. Located on Sunset Boulevard and designed by Philippe Starck to evoke a European chateau, XIV will feature Mina's modern American cuisine with a fluid, convivial atmosphere and unique menu structure.

"Michel Mina is an amazing chef as well as a longtime friend," said SBE CEO Sam Nazarian. "The stars have truly aligned to integrate Michael's culinary concept with the stunning design of Philippe Starck at this incredible location on Sunset Boulevard. XIV will offer Angelenos the chance to experience Mina's masterful craft right here in LA." Highlights from the Shellfish menu will include Chilled Maine Lobster with potatoes, celery, chestnuts, and truffle as well as Salt & Pepper Big Fin Squid with vermicelli, carrots, sprouts, and ginger. Garden Vegetable selections will capitalize on the seasonal California produce and range from Salad of Golden Beets with endive, mâche and a hazelnut vinaigrette to Sugar Pie Pumpkin Dumplings with stewed cherries and crispy sage. To finish the meal, diners can end with a selection of choices from the Chocolate, Fruits & Nuts menu by rising star pastry chef Jordan Kahn. Highlights include Liquid Shortbread with ricotta, absinthe and zucchini, White Chocolate Cube with pistachio, sake, and chrysanthemum, and Passion Fruit Curd with tomato, cashew and jasmine.

Spanning an entire city block on what is arguably the best real estate on the fashionable Sunset Strip, a 12-million-dollar renovation will completely transform the space with an entirely new modern exterior and an opulent, supremely Starck interior. Starck's interior design juxtaposes contemporary elements with traditional European accents. One wall of the main dining room is covered in stainless steel panels, while the other part of the dining space features fireplaces, bookshelves, hand-painted artwork and mirrors. Chocolate brown velvet draperies, swan sofas from Italy and an ornately carved ceiling add to the dramatic surroundings. To take full advantage of perfect L.A. evenings, Starck has created a lavish outdoor patio, complete with a large bar.

10月份，Michael Mina与SBE集团合作，在美国洛杉矶著名的时尚圣地日落大道开设了他的第一家洛杉矶餐厅。新餐厅还有一个与众不同的名字——XIV。XIV餐厅将厨师的饮食理念与设计师的设计构思完美融合，将给美国的饮食界带来前所未有的新体验。

SBE的首席执行官Sam Nazarian说：“Michael Mina既是一名优秀的厨师，也是我一生的挚友。XIV餐厅将使大家有机会在洛杉矶品尝Michael Mina的经典菜品。”

说到食物，这里的菜单不得不提。作为一个供应现代风格食物的餐厅，其所供应的食材琳琅满目，小到应季的蔬菜、水果，大到海鲜、禽肉，包括土豆、芹菜、栗子、龙虾、松露、椒盐大鳍鱿鱼、胡萝卜，豆芽、姜等等。用完正餐之后，客人可以选择性地品尝一下饮食界新星Jordan Kahn所制作的甜点以及饮品。重点推荐包括：奶酪脆饼、苦艾酒、美洲南瓜、开心果、日本清酒、菊花茶、腰果和茉莉花茶。

XIV餐厅位于时尚圣地——日落大道，12万美元的整修费用，会将其打造成一个奢华的餐厅。其内部装饰别具一格，设计师Philippe Starck以当代设计风格而著名，却赋予XIV餐厅以贵族气的欧式装饰，来到这里的客人宛如走进一座古老的欧洲贵族古堡，古典的豪华与精致会让人怀疑自己是否仍然身处21世纪。餐厅的墙面上铺置了不锈钢板，而其对面则是壁炉、书架、手绘艺术品和镜子。棕色天鹅绒窗帘、意大利进口沙发、雕刻华丽的天花板都赋予餐厅以高贵之感。为了能使客人更好地欣赏洛杉矶的夜景，设计师特意建造了一个带有大型酒吧的豪华室外庭院。

A VOCE

Designer: David Rockwell
Design Company: Rockwell Group
Location: New York, USA
Area: 836 m^2
Photographer: Bruce Buck
Portrait Photographer: Blandon Belushin

Rockwell Group opened up the existing space in the Time Warner building, so that guests would be aware of the park surroundings from every corner. The hues throughout are a contrast of ivory, chocolate and warm walnut wood. A modern Milanese aesthetic and allusions to tailored Italian fashion complement the fine Italian cuisine. The result is a rich, crafted, fashion-forward restaurant that will be both sleek and traditional.

We created a two-faceted entrance. The left side features a red backlit laser cut "A Voce" against a Calcutta marble wall. The right is a series of lit Kenon panels with a wood-grain texture to reference finely crafted Italian design. The bar is elevated to provide a theatrical platform from which to view the diners as well as the panoramic views around the main dining room.

The bar itself is adorned with a bamboo wood counter top, a Venetian patterned leather bar die, over marble flooring. A focal point of the space is a dramatically lit, temperature controlled glass wine case that stretches the length of the wall, separating the main dining room from the private and wine dining rooms. The mullions of this wine case wall are covered in stitched cognac-colored leather.

The dining tables have a modern chrome trumpet base with a chocolate-colored leather top and a walnut edge. They sit on walnut wood flooring, below a ceiling upholstered with a sleek Italian high-glossy white fabric, laid out in a herringbone pattern. These ceiling panels are interrupted by recessed chandeliers inspired by the mid-19th century Italian design, with interlaced antique bronze and blackened steel tubes and panels.

The private dining room is an extension of the main dining room that can serve a larger dinner party, or can be broken into two rooms with leather panels. Cutting-edge art work from a private collection adorns the walls, such as two portraits by Gary Hume.

The private dining room is surrounded by glass walls exhibiting over nine thousand of A Voce's wines. At the center of the space is a large table with a tooled walnut finish and a custom glass chandelier.

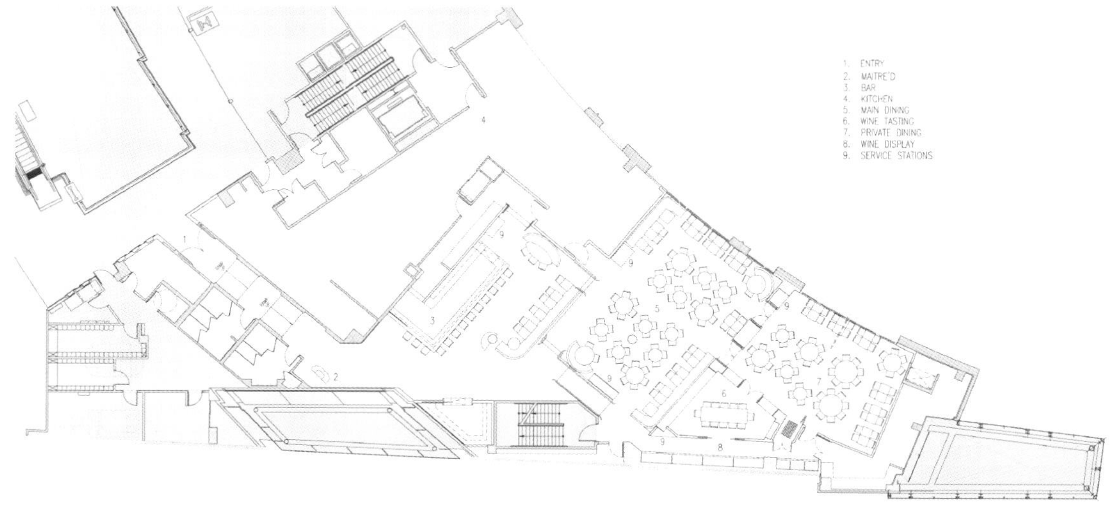

Rockwell集团拓宽了时代广场的现有空间，因此，客人可以从各个角度看到停车场及其周边环境。整体色调呈象牙色与巧克力色，与温暖的胡桃木色形成鲜明的对比。现代的米兰美学风格和意大利的时尚元素为该餐厅增色不少。这样的设计令一个富丽堂皇、精心雕饰、引领时尚、新潮而又传统的餐厅呈现在人们面前。

设计师打造了一个双重入口。悬挂在加尔各答大理石墙面上的红色激光背景灯是左侧入口的标志。右侧是经激光接口技术处理的板墙，精巧的纹理展现了精湛的意大利雕刻技艺。酒吧的设计高出地面，站在酒吧中间，可以看到所有的客人，将整个就餐区一览无余。竹饰的吧台和绘有威尼斯图饰的皮革在模具大理石地面的衬托下格外醒目。空间的焦点是一个炫丽的温控玻璃酒柜，背墙而立，一路延伸下去，将主餐区和包房区隔离开来。酒柜的竖框包裹着法国可那克颜色的手工皮饰。

餐桌的底座是时髦的铬质底座，呈喇叭形状。桌面是巧克力色的皮面和胡桃木框。它们交错摆放在胡桃木地板上，在意大利白色布艺的天花板下，尽显奢华。天花板上镶嵌着19世纪中期的意大利风格的枝形吊灯，古色古香的铜管和黑色的钢板错落有致。

包房区可以承办大型聚餐，也可以利用皮质屏风分隔成两个独立的包间，它是主餐区的延续。墙面上装饰着通过淘宝得到的最前卫的艺术品，例如，Gary Hume的两幅肖像画。这批淘品中的大多数在餐厅的其他功能区也随处可见。

该包房由玻璃墙围拢而成，墙上陈列着9 000瓶Voce酒。包房中心摆放着一张大餐桌，饰以做工精致的胡桃木装饰，餐桌上方悬挂着传统的枝形吊灯。

STIX RESTAURANT AND LOUNGE

Designer: 3six0 Architecture
Location: Boston, Massachusetts, USA
Client: Helvetica Group
Area: 139 m^2
Photographer: John Horner

Continuous band coils the space. On its way, the band folds up, out and over, as the wood wall panel, table, floor and ceiling. The experience is continuous from front to back, bar to restaurant and table to table. The bourbon toned wood, raw steel and cork present a simple unplugged elegance that is taut like a barrel. Tables fold up into the wall, enabling the space to transform into a cocktail party space, inspired by the food on a stick concept for the restaurant.

There were two main generative challenges inherent in the program. One was to create a sense of communal dining while maintaining a more private table configuration. The other was the need for two modes of operation — sit-down dining to standing room only. Given the lack of table storage, the design had to address these ambitions directly.

The solution to the challenges stated above was to create a continuous coil through the space. That continuity is interrupted when the mode is changed from sit-down dining to standing room only. The wood tables fold up into the wall panels clearing the floor and integrating into the graphic quality of the walls.

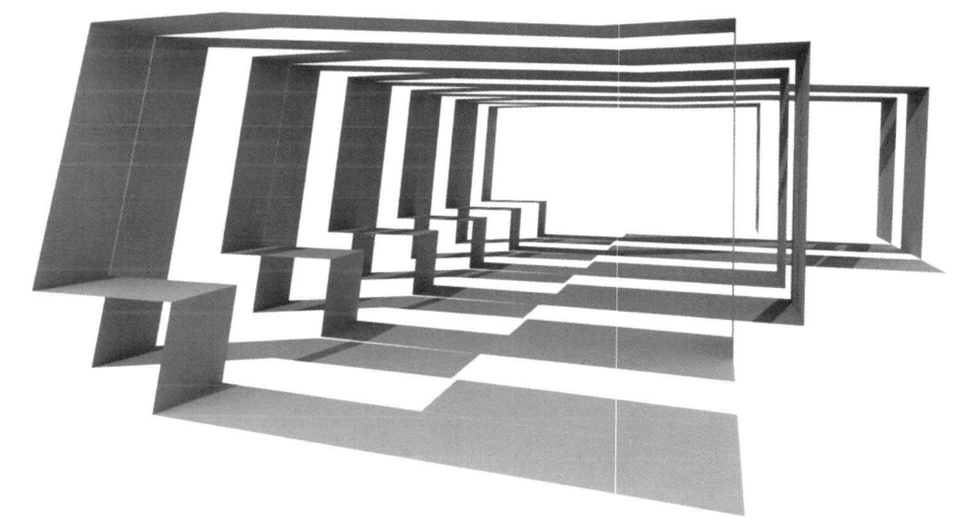

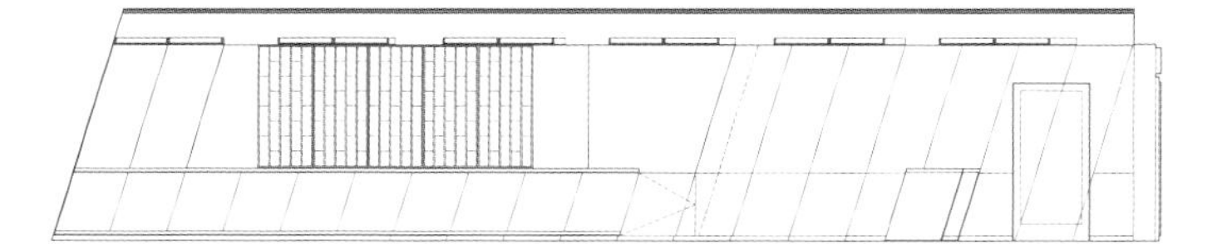

木质板材围拢成现在的空间。一路看下去，可见板材折折叠叠，突出又骤停，例如，镶嵌在木质墙面上的嵌板、餐桌、地板和天花板。从前到后，从酒吧到餐厅，甚至是在桌与桌的间隙中，木质板材都随处可见。波旁威士忌格调的木材、生铁和软木呈现出一种简约的优雅，好像一个啤酒桶。将餐桌嵌入墙体，就能呈现一个举办鸡尾酒会的空间，这个创意来自于人们希望在该餐厅享用串烧美食的真实想法。

该餐厅设计有两大挑战：一个是要营造出一种公共就餐场所的氛围，同时还要保持桌与桌之间良好的私密性；另一个是为了满足两种就餐模式的需求——店内就餐和外卖窗口。如果没有足够的餐桌，这个设计就非它莫属了。

解决上述问题的办法是营造一个环形就餐区，这个环形只是在需要设置店内就餐和外卖时才被打断。木质餐桌可以折叠起来嵌入墙板，既节约了空间，又与墙面巧妙地融为一体。

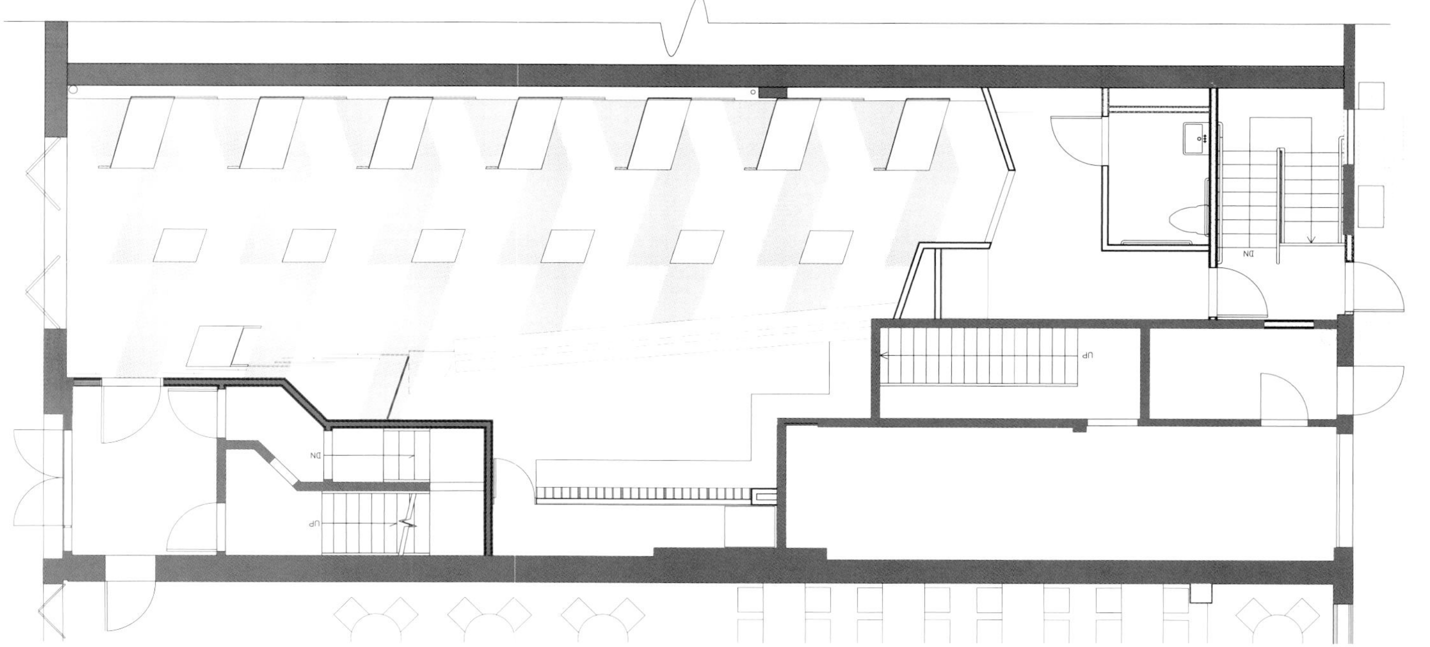

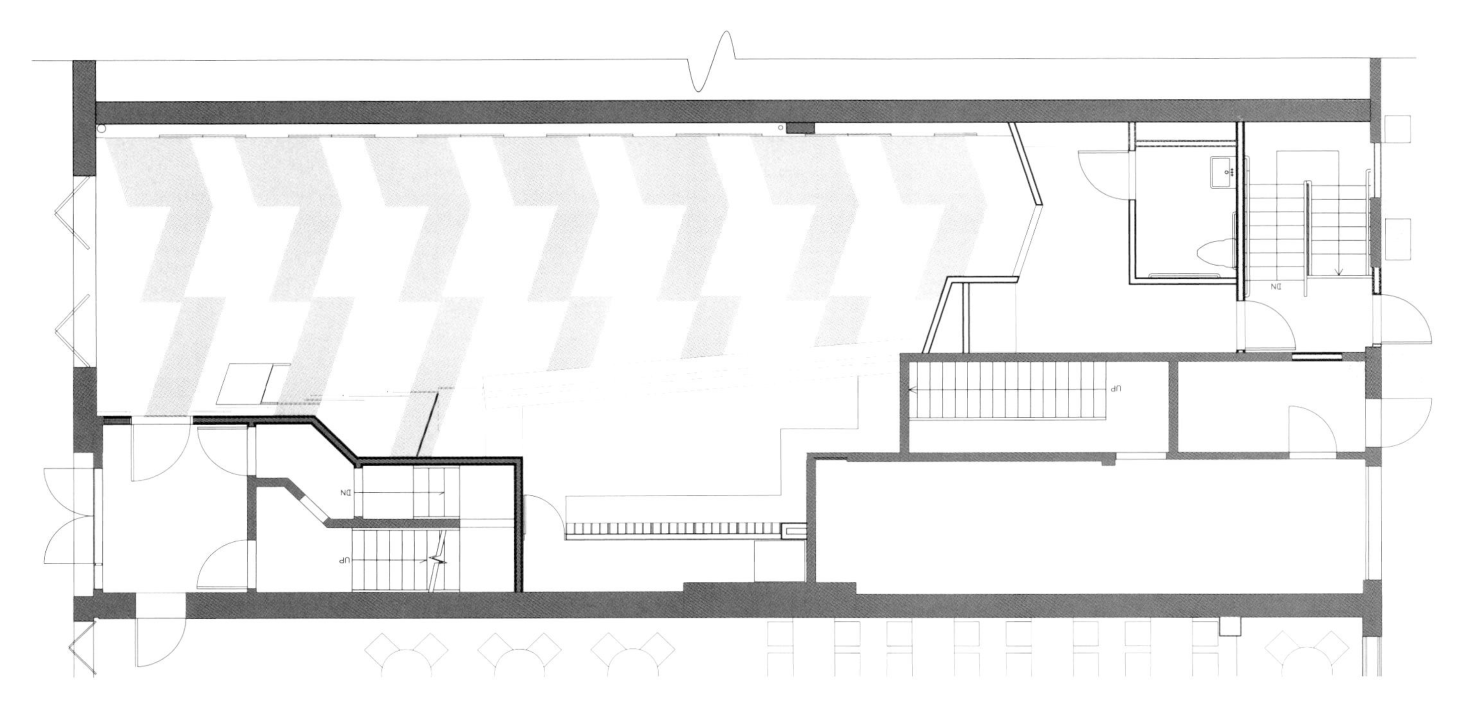
DN
UP
UP
DN
UP

CGV GOLD CLASS LOUNGE

Designer: Younjin Jeong, Daesoon Hwang
Design Company: Urbantainer Co., Ltd.
Area: 253.8 m^2
Location: 7F, I Park Mall, 40-999, Hangangno 3-ga, Yongsan-gu, Seoul, Korea
Photographer: Namgoong Sun, Urbantainer

CGV Yongsan Gold Class Lounge is the space to be provided to the CGV theater users. It is made of three different themed spaces that address specific needs to three different client targets. Café Zone is accessible to all incoming CGV theater users. Gold Zone is accessible to all the Gold Class clients. Private Zone is the celebrity and VIP space for publicity events and junkets. These three spaces are divided specifically according to the thematic materials used for each individual space. The designers emphasized the creative direction on to the exterior design. Exterior design is the first thing that the CGV users notice as they climb up the escalator to get to the entrance of the theater which is where the designers wanted to deliver the most impactful space design concept for all too see. The former Yongsan CGV Theater was initially designed when big multiplexes were in demand therefore materials, for example, neon's and metals were mainly used to purvey the cinematic fantasy concept in a forceful style. Redesigning and replacing the circular structure on the exterior space which was connected to the previously existing ceiling was the main concern, especially considering the given budget and the construction process schedule. The designers tried to maintain the former structure, yet by incorporating the pattern and combination of finishing materials, the designers were able to create the dynamic rhythm for the space. This concept mirrors the vitality of Yongsang CGV which draws the most theatergoers out of all the CGV multiplexes in Seoul. By simplifying to the maximum the types of finishing materials used and designing the patterns through maintaining the former structure, the designers were able to pertain the edginess and young energy of younger theatergoers who are the major clientele of CGV Yongsan. The exterior design is the crucial space where the guests can feel this concept of dynamic rhythm the most. The guests who climb up the escalator can fully sense the tension of rhythm emanating from the pattern of finishing materials. Utilizing the vertical repetition of the pattern without destroying the wood material plays up the characteristics of the site. The atmosphere of the interior is that of a cozy cafe where the designers used the raw interior concrete block and UB block, yet by applying the patterns from maintaining the structure of the finishing materials, the designers were able to convey the feeling of liveliness.

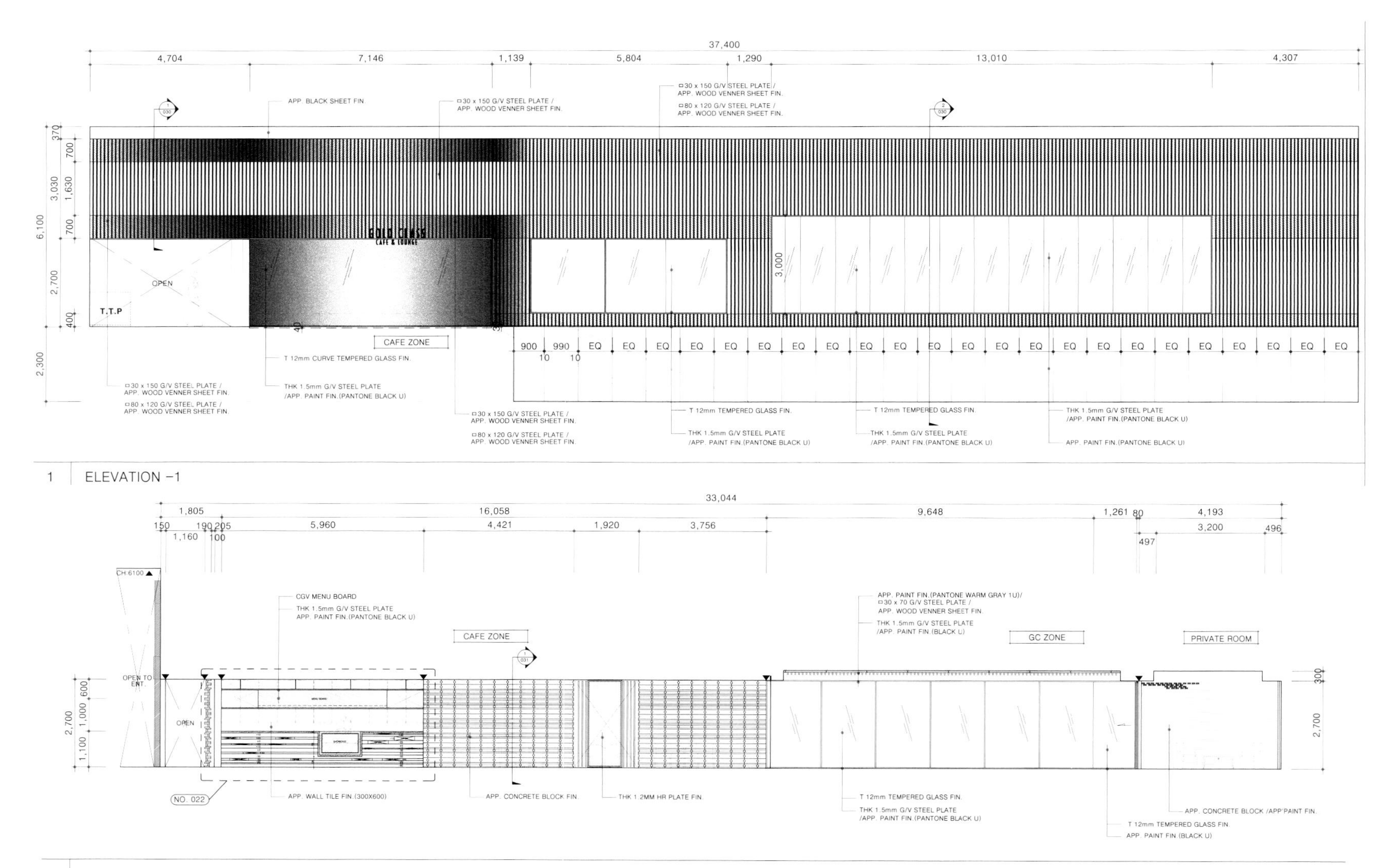
37,400
4,704
7,146
1,139
5,804
1,290
13,010
4,307
APP. BLACK SHEET FIN.
□30 x 150 G/V STEEL PLATE / APP. WOOD VENNER SHEET FIN.
□30 x 150 G/V STEEL PLATE / APP. WOOD VENNER SHEET FIN.
□80 x 120 G/V STEEL PLATE / APP. WOOD VENNER SHEET FIN.
GOLD CLASS
CAFE & LOUNGE
OPEN
T.T.P
CAFE ZONE
T 12mm CURVE TEMPERED GLASS FIN.
□30 x 150 G/V STEEL PLATE / APP. WOOD VENNER SHEET FIN.
□80 x 120 G/V STEEL PLATE / APP. WOOD VENNER SHEET FIN.
THK 1.5mm G/V STEEL PLATE /APP. PAINT FIN.(PANTONE BLACK U)
□30 x 150 G/V STEEL PLATE / APP. WOOD VENNER SHEET FIN.
□80 x 120 G/V STEEL PLATE / APP. WOOD VENNER SHEET FIN.
T 12mm TEMPERED GLASS FIN.
THK 1.5mm G/V STEEL PLATE /APP. PAINT FIN.(PANTONE BLACK U)
T 12mm TEMPERED GLASS FIN.
THK 1.5mm G/V STEEL PLATE /APP. PAINT FIN.(PANTONE BLACK U)
THK 1.5mm G/V STEEL PLATE /APP. PAINT FIN.(PANTONE BLACK U)
APP. PAINT FIN.(PANTONE BLACK U)
1 ELEVATION -1
33,044
1,805
16,058
9,648
1,261
4,193
CGV MENU BOARD
THK 1.5mm G/V STEEL PLATE APP. PAINT FIN.(PANTONE BLACK U)
CAFE ZONE
APP. PAINT FIN.(PANTONE WARM GRAY 1U)/ □30 x 70 G/V STEEL PLATE / APP. WOOD VENNER SHEET FIN.
THK 1.5mm G/V STEEL PLATE /APP. PAINT FIN.(BLACK U)
GC ZONE
PRIVATE ROOM
OPEN TO ENT
OPEN
NO. 022
APP. WALL TILE FIN.(300X600)
APP. CONCRETE BLOCK FIN.
THK 1.2MM HR PLATE FIN.
T 12mm TEMPERED GLASS FIN.
THK 1.5mm G/V STEEL PLATE /APP. PAINT FIN.(PANTONE BLACK U)
APP. CONCRETE BLOCK /APP.PAINT FIN.
T 12mm TEMPERED GLASS FIN.
APP. PAINT FIN.(BLACK U)
1 ELEVATION -1

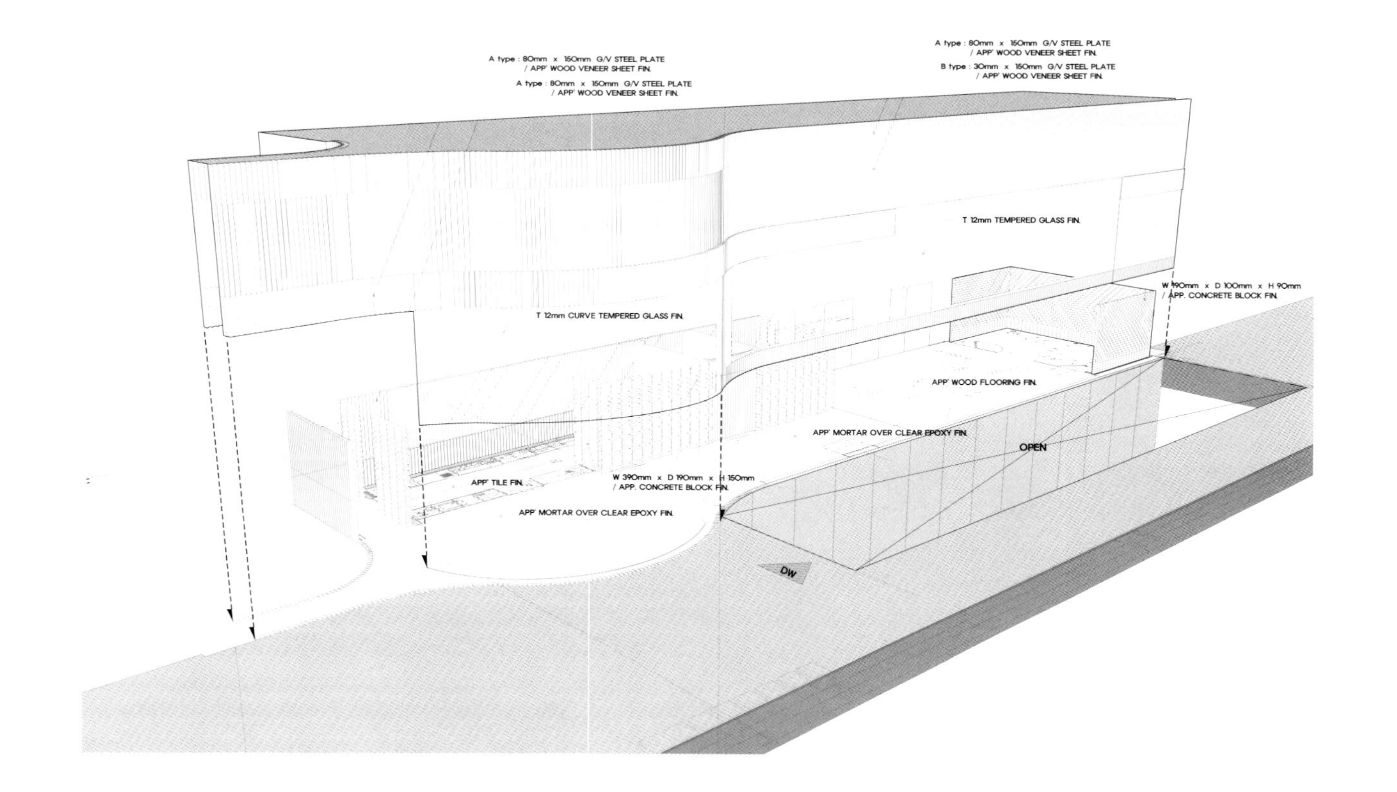
A type : 80mm x 160mm G/V STEEL PLATE / APP' WOOD VENEER SHEET FIN.
A type : 80mm x 160mm G/V STEEL PLATE / APP' WOOD VENEER SHEET FIN.
A type : 80mm x 160mm G/V STEEL PLATE / APP' WOOD VENEER SHEET FIN.
B type : 30mm x 160mm G/V STEEL PLATE / APP' WOOD VENEER SHEET FIN.
T 12mm TEMPERED GLASS FIN.
T 12mm CURVE TEMPERED GLASS FIN.
APP' WOOD FLOORING FIN.
APP' MORTAR OVER CLEAR EPOXY FIN.
OPEN
APP' TILE FIN.
W 390mm x D 190mm x H 150mm / APP. CONCRETE BLOCK FIN.
APP' MORTAR OVER CLEAR EPOXY FIN.
DW

GOLD CLASS
CAFE
LOUNGE

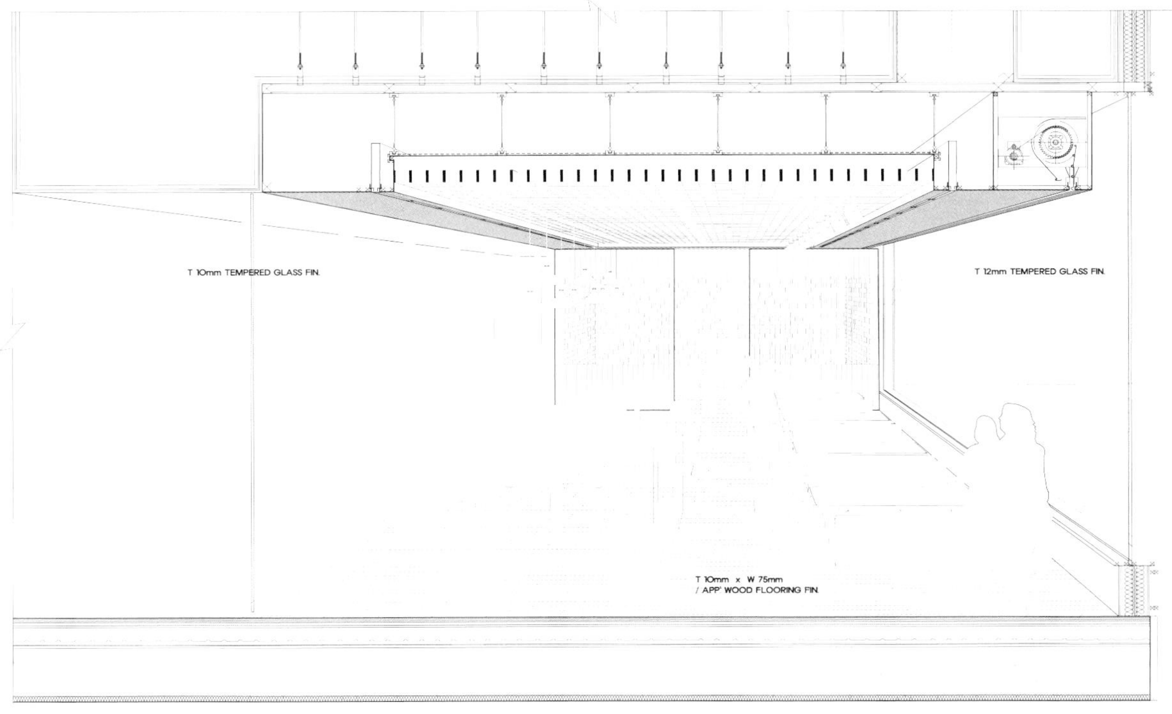
T 10mm TEMPERED GLASS FIN.
T 12mm TEMPERED GLASS FIN.
T 10mm x W 75mm
/ APP' WOOD FLOORING FIN.

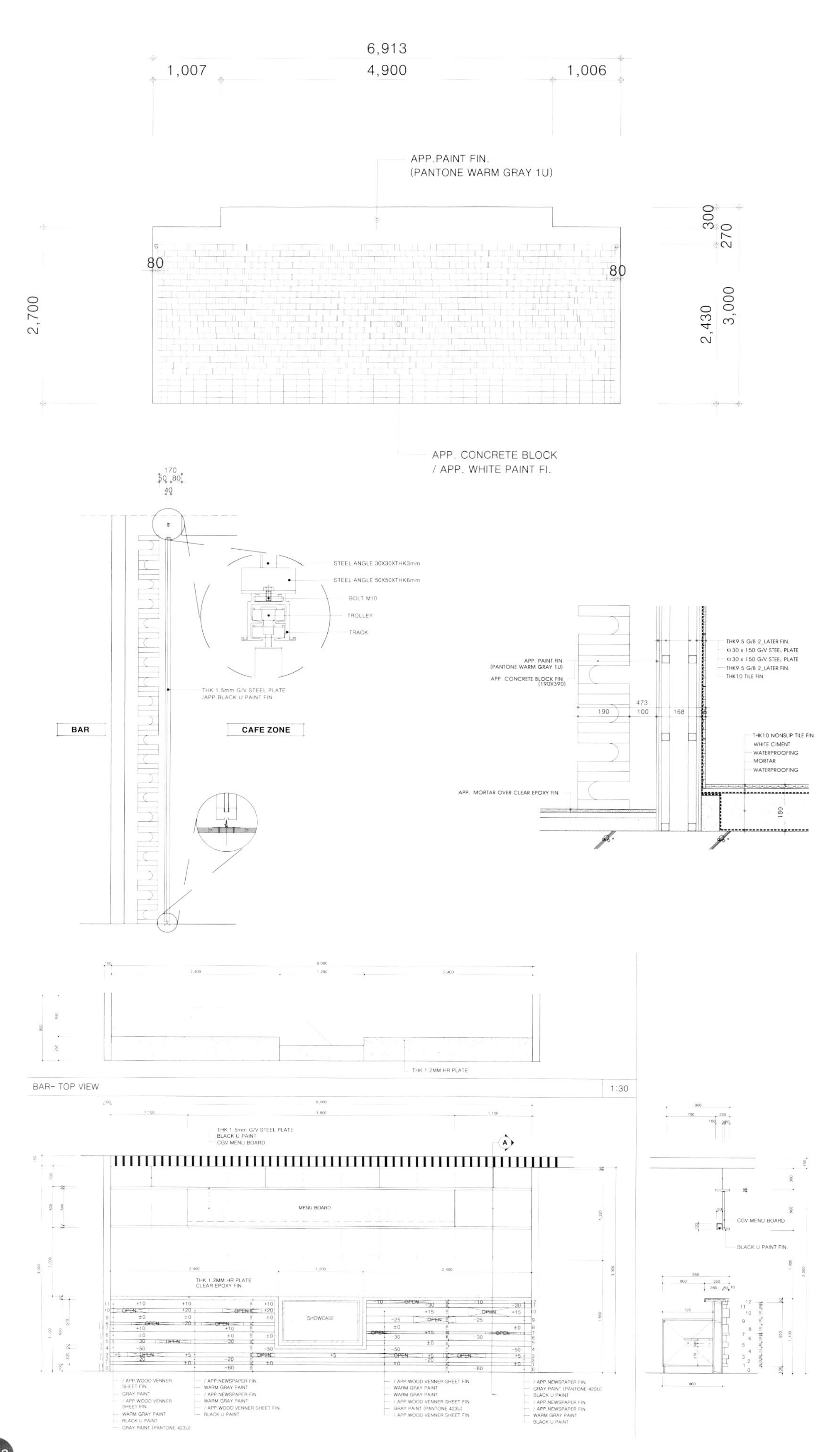

　　韩国Yongsan CGV株式会社金卡会客厅是为CGV剧院观众提供服务的场所。主要有三个不同的主题空间，解决三种不同客人的需求。咖啡厅是为所有的CGV剧院观众提供服务的，金卡区是为所有金卡客户服务的，私人区是为名人和VIP客户举行公共活动和宴会的场所。根据每个独立空间所使用主题材料的不同，这三个空间被单独隔离开来。在外观设计上，突出了设计师的创意理念。CGV观众乘电梯来到剧院入口，首先看到的就是外部设计，这也是想向观众传递的最具有视觉冲击力的设计。当对原来的汉城CGV剧院进行设计时，正需要多个放映厅，因此主要使用了霓虹灯和金属这样的材料，以强有力的方式打造出电影院的梦幻概念。外面的圆形结构，是和原来的天花板相连接的，现在要替换掉并重新进行设计，因为有预算和改造日程表的限制，这成为设计师最担心的事情。他们通过采用一些图案和涂饰材料，在尽力保留原先结构的前提下，打造出空间的动态感。这个理念反映出了 Yongsang CGV的活力，能够吸引首尔更多的观众前来参观。将使用的涂饰材料种类简单化，并且通过保留原来结构，设计出新的图案，既迎合年轻观众，又体现他们的活力。因为他们是CGV Yong San的主要观众群。外部设计是观众最能感知这个空间动态感的地方。乘电梯上来的观众能够完全感受到从涂饰材料中散发出来的节奏感。不破坏木质材料，合理利用图案的纵向重复，渲染出该场所的特点。内部的咖啡厅采用的是未经加工过的混凝土方块和UB方块，保留涂饰材料的结构，制造出活力十足的景象。

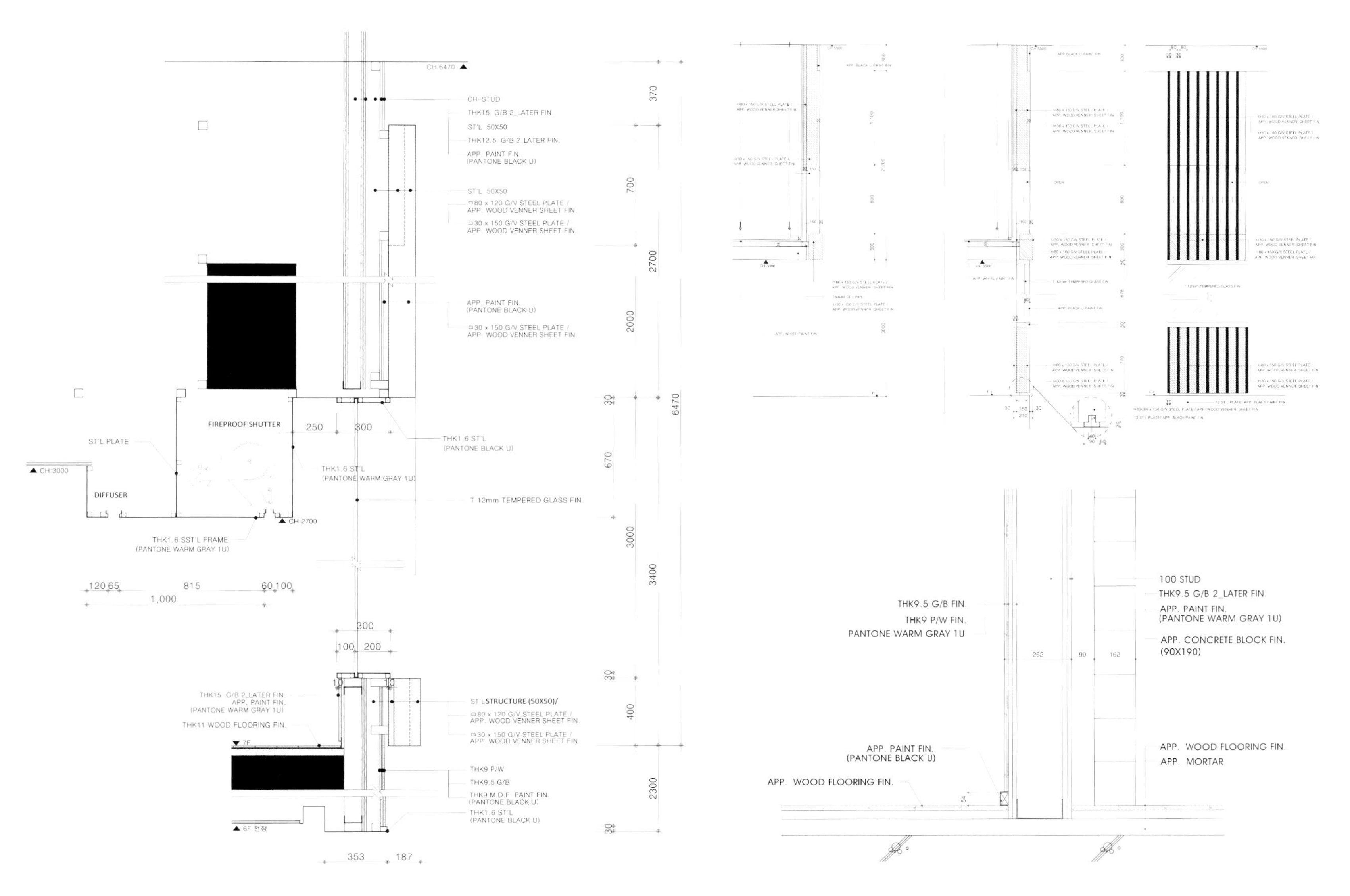
CH 6470
CH-STUD
THK15 G/B 2_LATER FIN.
ST'L 50X50
THK12.5 G/B 2_LATER FIN.
APP. PAINT FIN.
(PANTONE BLACK U)
ST'L 50X50
□80 x 120 G/V STEEL PLATE /
APP. WOOD VENNER SHEET FIN.
□30 x 150 G/V STEEL PLATE /
APP. WOOD VENNER SHEET FIN.
FIREPROOF SHUTTER
ST'L PLATE
CH 3000
DIFFUSER
THK1.6 ST'L
(PANTONE BLACK U)
THK1.6 ST'L
(PANTONE WARM GRAY 1U)
T 12mm TEMPERED GLASS FIN.
CH 2700
THK1.6 SST'L FRAME
(PANTONE WARM GRAY 1U)
THK15 G/B 2_LATER FIN.
APP. PAINT FIN.
(PANTONE WARM GRAY 1U)
THK11 WOOD FLOORING FIN.
ST'L STRUCTURE (50X50)/
THK9 P/W
THK9.5 G/B
THK9 M.D.F. PAINT FIN.
(PANTONE BLACK U)
6F 천장
100 STUD
THK9.5 G/B 2_LATER FIN.
APP. PAINT FIN.
(PANTONE WARM GRAY 1U)
APP. CONCRETE BLOCK FIN.
(90X190)
THK9.5 G/B FIN.
THK9 P/W FIN.
PANTONE WARM GRAY 1U
APP. PAINT FIN.
(PANTONE BLACK U)
APP. WOOD FLOORING FIN.
APP. WOOD FLOORING FIN.
APP. MORTAR

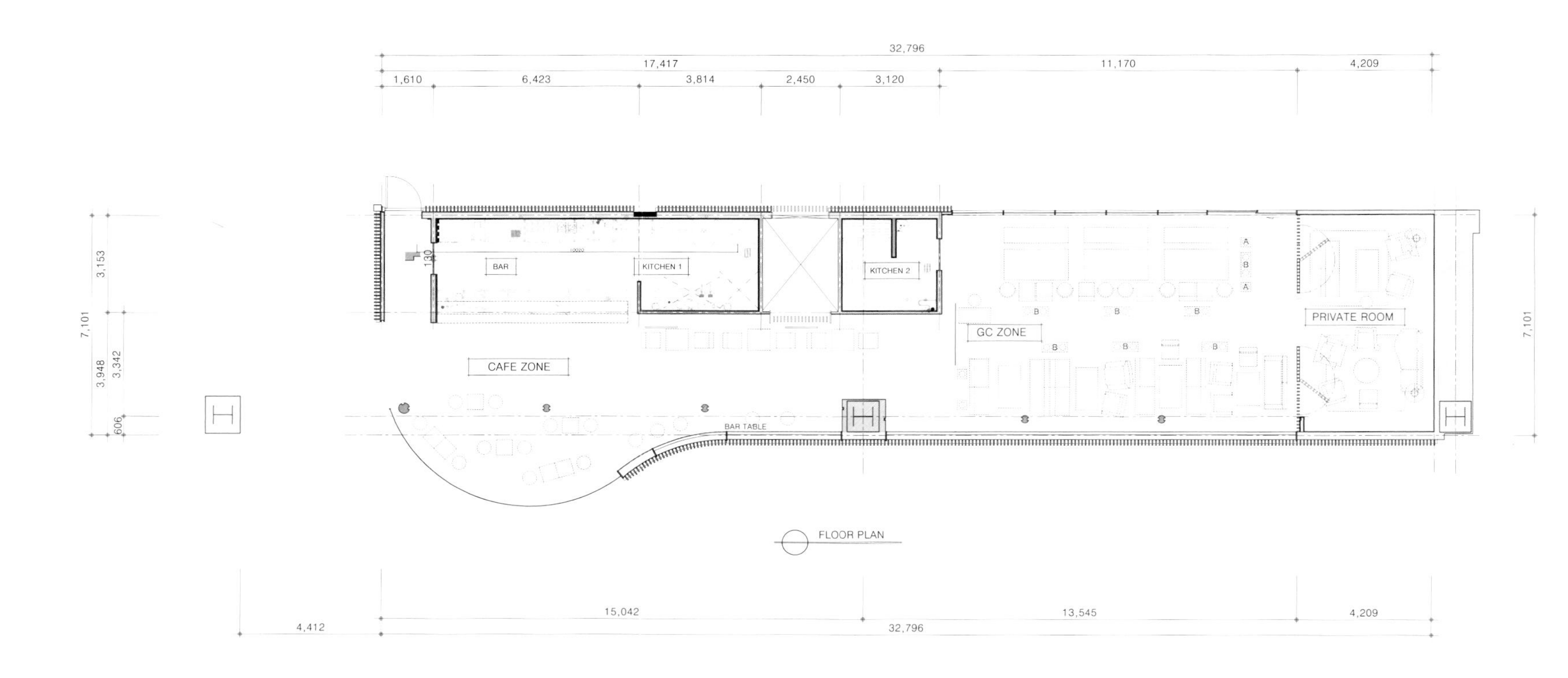
32,796
17,417
11,170
4,209
1,610
6,423
3,814
2,450
3,120
BAR
KITCHEN 1
KITCHEN 2
GC ZONE
PRIVATE ROOM
CAFE ZONE
BAR TABLE
FLOOR PLAN
3,153
7,101
3,948
3,342
606
4,412
15,042
13,545

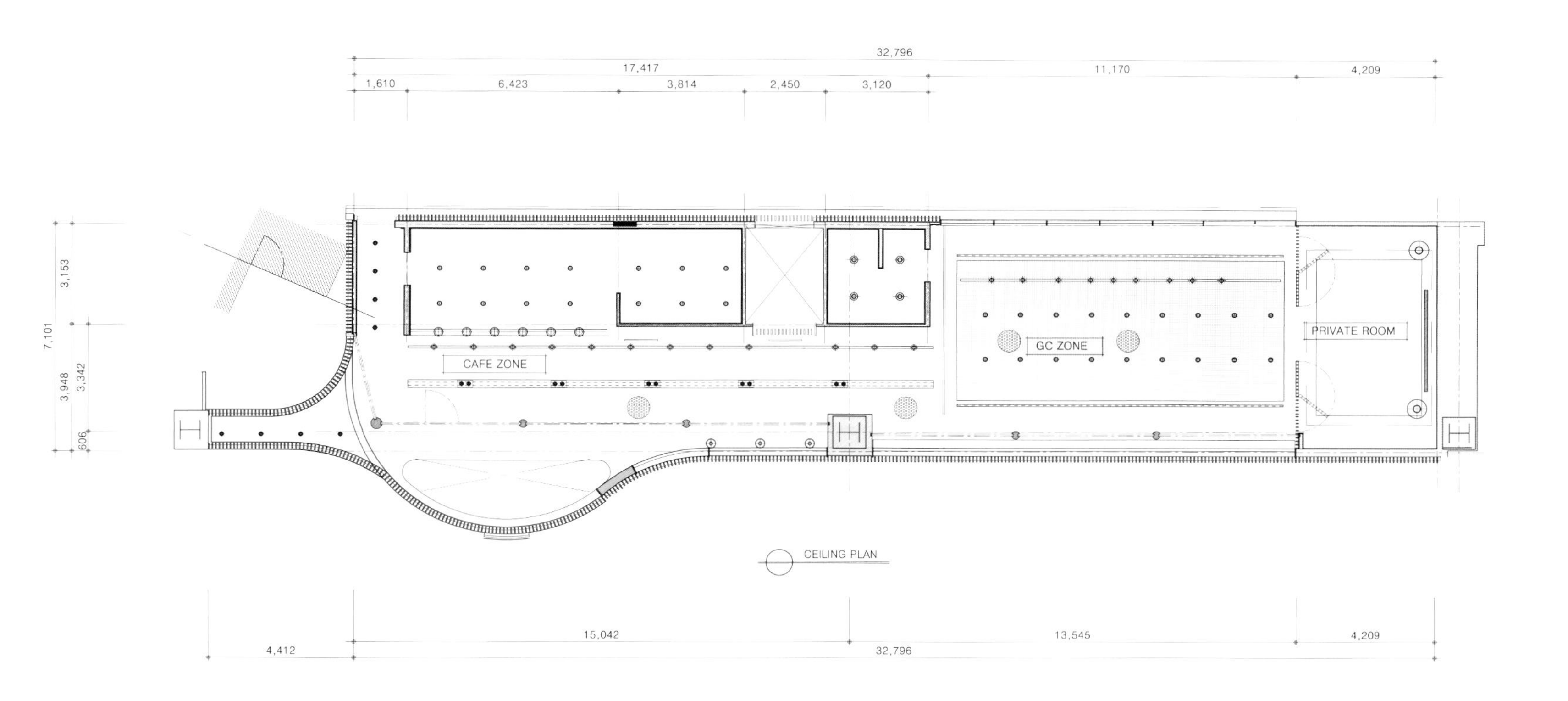
32,796
17,417
11,170
4,209
1,610
6,423
3,814
2,450
3,120
3,153
7,101
3,948
3,342
606
CAFE ZONE
GC ZONE
PRIVATE ROOM
CEILING PLAN
15,042
13,545
4,209
4,412
32,796

LA GRELHA RESTAURANT

Designers: Jorge Luis Hernández Silva, Diego Vergara, Lorena Ochoa
Firm: Hernandez Silva Arquitectos
Location: Guadalajara, Jalisco, Mexico
Photographer: Carlos Díaz Corona

The gastronomic concept of the restaurant comes from its name. "La Grelha" which means barbecue in Portuguese, where all the formal concept of the project is created. The space is a high terrace on which fine columns stand and are topped by a nearly flat steel structure, and covered with wood that wraps and shapes the space. The walls keep the brick in a raw manner only painted in white, and are only used for the service area. A very thick wall creates a boundary and marks the closing of the street where the logo of the restaurant is also placed. The entrance is confined to the corner, which welcomes guests and gives a very nice waiting room. This space also is bounded to the bottom of a metallic double curtain where vegetation lines climb to make a big green box that lights up at night and in the evening by the setting sun. There is also a wall coming down from the roof and floats above the water surface.

The interior space is clear and very bright, due to space where light interpass the ceiling. The bar is at the center and it is defined by a squared lattice. It divides a little the space into two, allowing one part to look over street to the east and the other to the north where big trees in the side street can be seen. This area is elevated to trick the car parking, and the space is prepared to extend a wooden platform area over the cars and to define an area for smokers. Both spaces are linked by a very long reflecting pool with wall of climbing plants to give depth to the space. At the background, guests can observe huge grills with their extractors.

The whole place was built on a short budget, so the designers opted for the use of simple materials. Even the red glass lighting is very characteristic of the region, and the floor is quarry and most of the furniture is pine.

Guadalajara has a perfect climate almost all year, so making a terrace is ideal for this food concept. Rain can be sometimes heavy but for very short time; here, it can rain in one hour what falls in a city for a year. Therefore, the terrace has a hidden window behind wooden marimbas and only extends when the weather demands it.

LA GRELHA

LA GRELHA

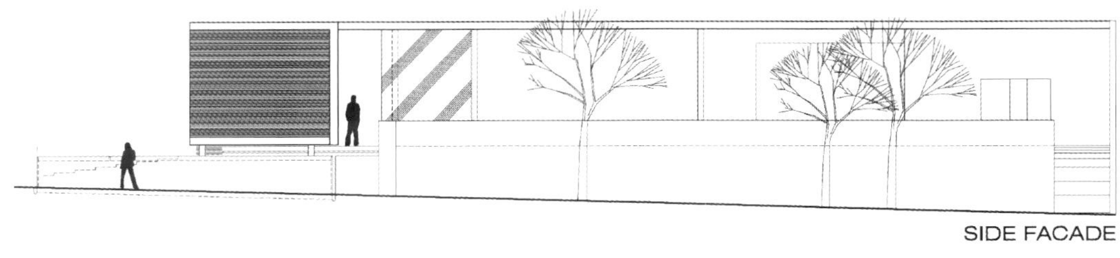

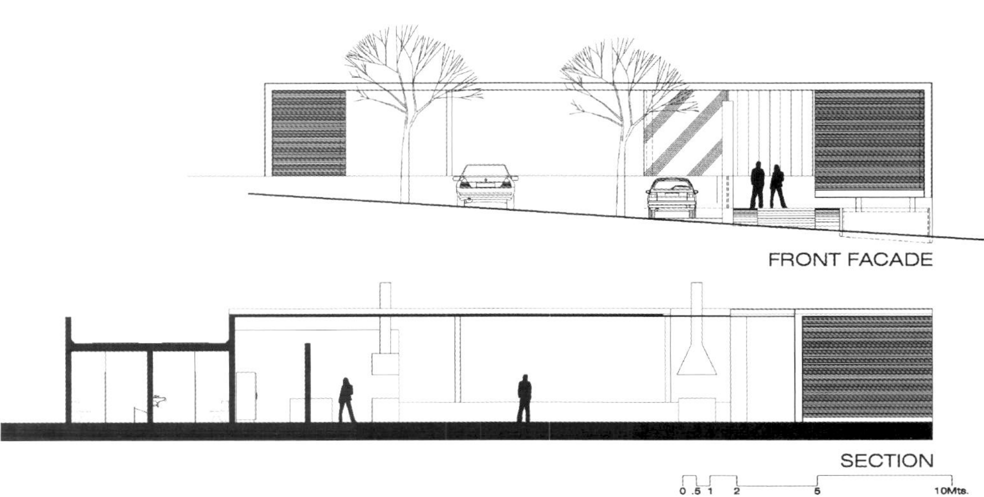

餐厅的烹饪理念源于它的名字La Grelha，在葡萄牙语里是烧烤的意思，这正是该项目创建的基础。整个空间是一个高高的露台，矗立着几根柱子，露台顶上搭建了一个近乎平整的钢结构，另外用木材覆盖并打造了整个餐厅。墙壁也保留了砖的本色，只简单地涂刷了一层白色，并且只用于服务区。一堵厚墙分出了餐厅和街道的边界，餐厅的标志正印刷在这面墙上，入口处也位于这面墙的一角，为客人营造了一个舒服的候餐室；这个候餐室的另外一面是双层的金属幕帘，自然植被可以沿着攀爬上去，形成了一个大的绿色盒子，傍晚时落日可以照到这个区域，夜晚时灯光可以用来照亮。另外，还有一面墙从顶棚延伸下来，一直到达其下方的水平面上。

内部空间简洁、明亮。吧台位于中央，用正方形方格装饰的吧台将空间分成了两部分，一部分可以看到东边的街道，另一部分可以看到北边街旁的大树。这个区域可以巧妙地设计在停车场的上方，并且也可以在停车场的上面扩充一个木质平台，用做吸烟区。两部分由一个长长的倒影池连接起来，倒影池旁边的墙上爬满了植被，营造出纵深的感觉。在墙后面，可以看到巨大的带有提取器的烧烤架。

这个餐厅的预算很有限，所以设计师选择使用简单的材料。红色的玻璃灯是餐厅的特色，地板是方形石，大多数家具是松木制成的。

Guadalajara全年气候宜人，所以搭建一个露天平台是非常理想的，很适合烧烤的烹饪理念。有时也会下很大的雨，但是历时都很短，有时一年一个城市也就下一个小时的雨。因此露台也安装了隐形的窗户，在天气需要的时候可以伸展出来。

LA GRELHA

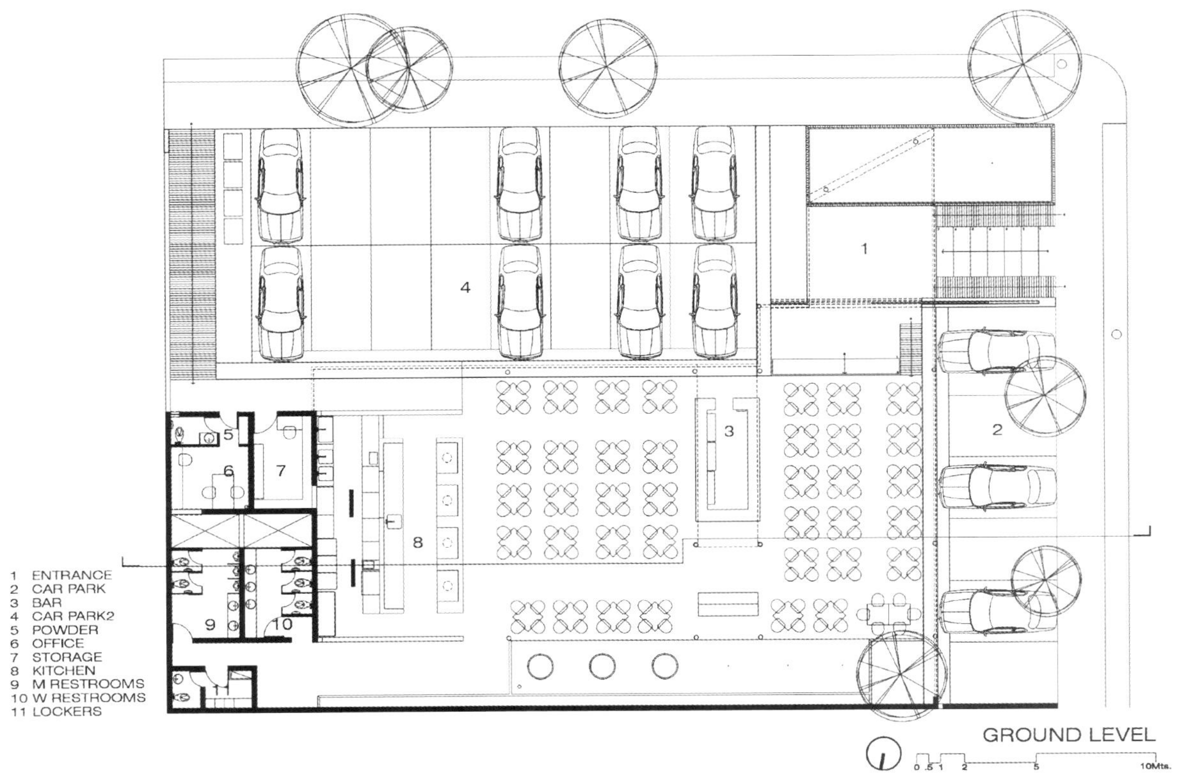
1 ENTRANCE
2 CAR PARK
3 BAR
4 CAR PARK2
5 POWDER
6 OFFICE
7 STORAGE
8 KITCHEN
9 M RESTROOMS
10 W RESTROOMS
11 LOCKERS
GROUND LEVEL
10Mts.

SANTA MARTA RESTAURANT

Designers: Studio Kuadra
Design Company: Studio Kuadra
Location: Mazze, Italy
Area: 110 m^2 Each Floor
Photographs: Photophilla

The chapel, located in the historical quarter of Mazzè, is part of the religious complex of the parish church built in the 18th century. The project consisted of the restoration of the chapel and its conversion into a restaurant. The building has three floors.
The ground floor was separate from the chapel above and used for storage. Now it is the main entrance to the restaurant; there is a bar, the kitchen and the toilets. The first floor (the old chapel) has been transformed into the main dining room. An interior gallery overlooks the dining room. This features the original double height vaulted ceiling. The gallery framework is in steel with wood flooring. To conserve the ancient chapel, the gallery is fixed to the old walls at only four points. The dining room can be seen through two openings which give onto small balconies. The balustrade is composed of glass and fixed to the floor. The stairs and gallery structure is made of metal painted white, and floor furniture and tables of wedge.
The stairs and gallery frameworks, some parts of furniture (bar counter, wine rack and table structure) are realized in metal strips painted white, this feature characterizes the restaurant.
The project aims to be linear and minimal; the only ornamental element is a floreal pattern, protagonist in the dining room: white petals that become wine racks or decoration for the glass wall of the office. The same floral pattern is used for the restaurant logo.

section

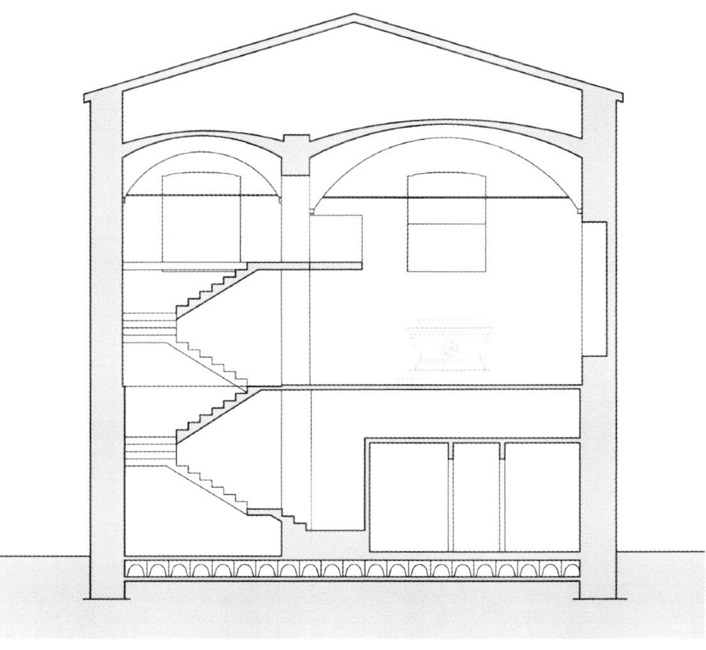

stair section

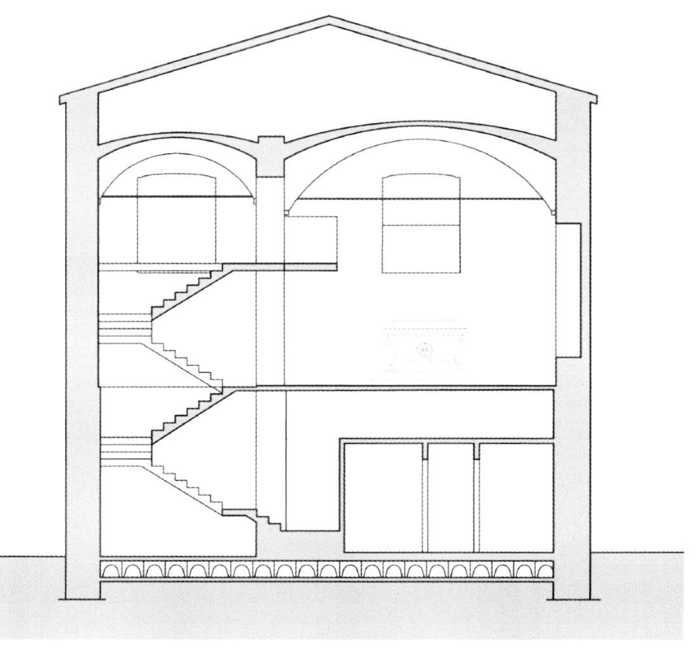

stair section

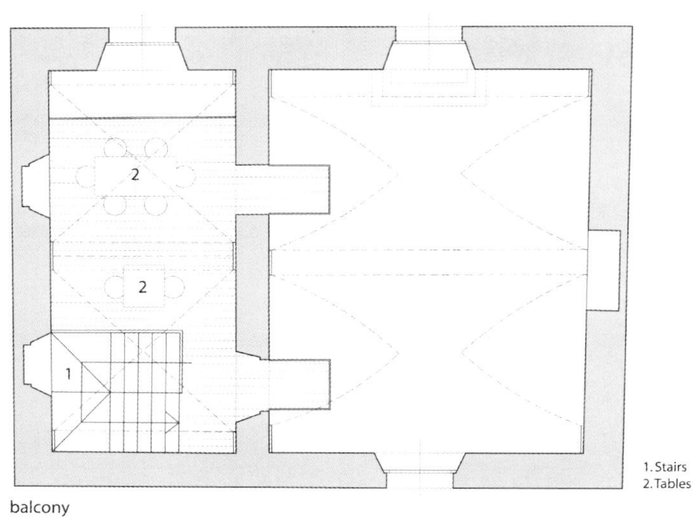
2
2
1
balcony
1. Stairs
2. Tables

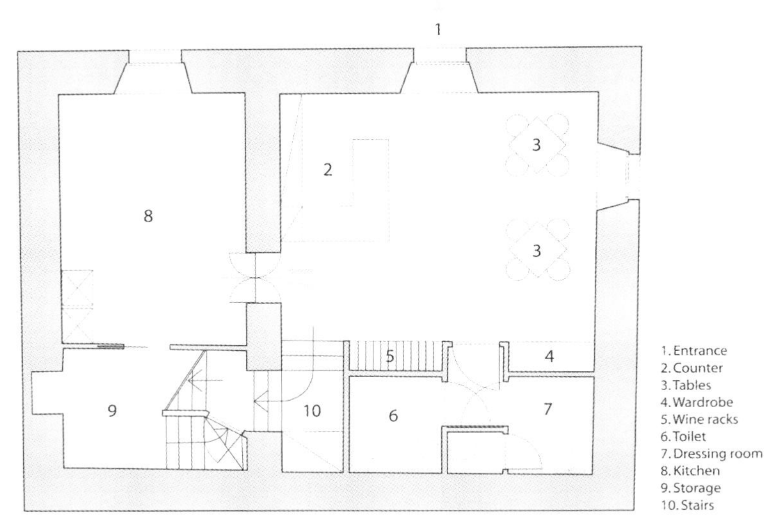

1. Entrance
2. Counter
3. Tables
4. Wardrobe
5. Wine racks
6. Toilet
7. Dressing room
8. Kitchen
9. Storage
10. Stairs

ground floor

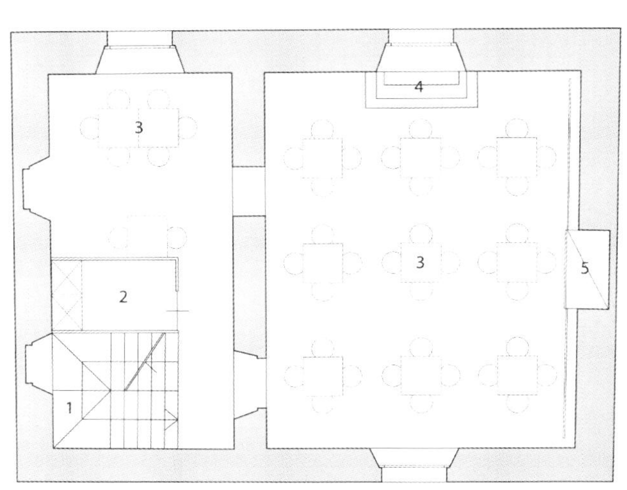

1. Stairs
2. Office
3. Tables (46 places)
4. Altar
5. Winw racks

first floor

sm

sm

该教堂位于Mazz的历史中心，是18世纪教区宗教建筑群的一部分。该项目包括对教堂的修复以及将其改建成餐厅。该建筑共有三层。

一层与上面的教堂相分离，用做储物室。现在用来作为餐厅的入口，这一层有吧台、厨房以及卫生间。二层（原教堂）被改建成餐厅。里面有室内装饰画廊。这使得拱形天花板变为原来的两倍高。画廊内的框架是钢制的，而地板是木质的。为了保留古式教堂的风格，画廊被固定在墙上的四个点上。阳台上的栏杆是由玻璃制成的，并且固定在地面上。客人站在户外阳台上，可以将餐厅的景色一览无遗。

楼梯、画廊结构以及餐厅内部的家具（吧台、桌椅）都是白色铁制的，这一切都成为该餐厅的显著特点。

餐厅内唯一的装饰元素就是花瓣图案。这些白色的花瓣，在酒架上或玻璃墙上随处可见。同样的花瓣图案也被设计师用来作为装饰餐厅的标志。

TWISTER RESTAURANT

Designers: Sergey Makhno, Butenko Vasiliy
Design Company: Sergey Makhno, Butenko Vasiliy
Location: Kyiv, Ukraine
Area: 421 m^2
Materials: Wood, Concrete, Metal, Marble, Plastic

Design by Serghii Makhno and Vasiliy Butenko, this is amazing interior for a restaurant in Kiev, where guests can fell like baby bird while drinking cocktail or have a dinner at tornado top. This restaurant can be classed as modern European and offers a molecular kitchen style dishes. The main aim while designing this restaurant space was to create an environment that is natural, modern and comfortable. This restaurant features two areas: a two-storeyed dining section and relaxing bar area.

Two-storeyed dining section was inspired by two natural phenomenas: tornado and rain. The space features six tornado shape balconies which create one dynamical upper zone with five dining cells. Restaurant walls lined with wooden slats create contrast balconies smooth surface. Ceiling lamps imitate rain drops falling from the sky so complete atmosphere is very natural and ensures comfort. Spaces of the restaurant are calming, due to the natural tones which extend throughout the restaurant : beige, ochre, garnet and brown.

dining area
wc
wc
kitchen area
wardrobe
bar area
office

该餐厅位于Kiev，由Serghii Makhno和Vasiliy Butenko进行设计，内部装修漂亮、惊人。在那里，客人感觉如同一个幼鸟，冒着飓风，喝着鸡尾酒或者吃着晚餐。这个餐厅是现代欧式建筑风格，并引入当下流行的分子料理。设计的主要目标是创建自然、现代和舒适的环境。餐厅有两个区域：一个两层的就餐区和一个放松的酒吧区域。

双层就餐区的设计是受两种自然现象启发的：龙卷风和雨。餐厅有六个龙卷风形状的露台，打造出有五个就餐包间的动态上层区域。餐厅的墙体上覆盖着木质板条，与露台的光滑表面形成对比。天花板的吊灯模仿雨滴从天空落下的形状，所以，整体氛围显现得非常自然和舒适。餐厅运用灰棕色、黄土色、深红色和棕色等色彩表达出的自然基调，凸显空间的安静。

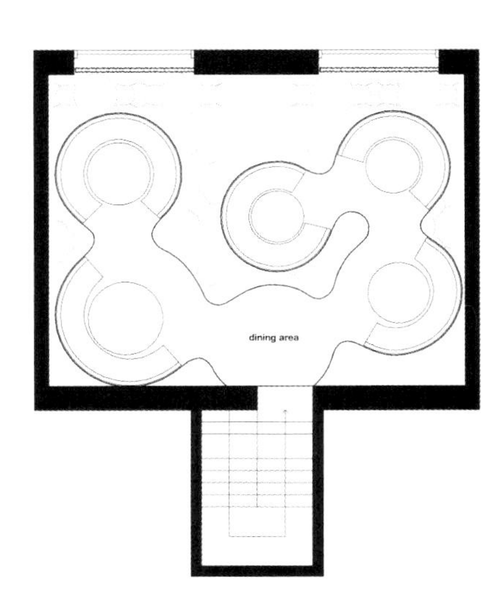

WEXLER'S RESTAURANT

Designers: Joshua Aidlin, David Darling, Roslyn Cole, Adrienne Swiatocha, Shane Curnyn
Design Company: Aidlin Darling Design
Location: 568 Sacramento Street, San Francisco, CA, USA
Area: 121 m^2
Photographer: Matthew Millman

Aidlin Darling Design designed the interior of Wexler's Restaurant composed of a restrained palette of zinc and rift-sawn oak warmed by the honey glow of indirect lighting.
Reminiscent of an agrarian heritage, the materials are given an uncharacteristic sense of refinement similar to the cuisine itself. Evocative of smoke and charred wood, a dynamic undulating ceiling installation composed of laser cut MDF fins bridges the gap between old and new by linking the modern insertion to the historic facade.
The solution consists of four primary elements that combine to create a warm and light filled restaurant that performs as a neutral backdrop in which the food can take center stage, but also serves as a narrative to highlight the innovative treatment of traditional barbeque cuisine.
Inside, the restrained palette of rift-sawn oak and zinc are used to pay homage to the agrarian and pastoral roots of barbeque. These traditionally utilitarian materials are treated with an uncharacteristic level of refinement analogous to that of the menu choices that are so atypical of traditional barbeque. In a similar vein, a traditional wrought iron chandelier is transformed into a modern sculptural showpiece with the application of blood red paint. The combination of warm, honey toned lighting and the specificity of the architectural language brings the palette out of the agrarian language and into an intimate environment appropriate to serve Wexler's impeccably prepared cuisine.

568

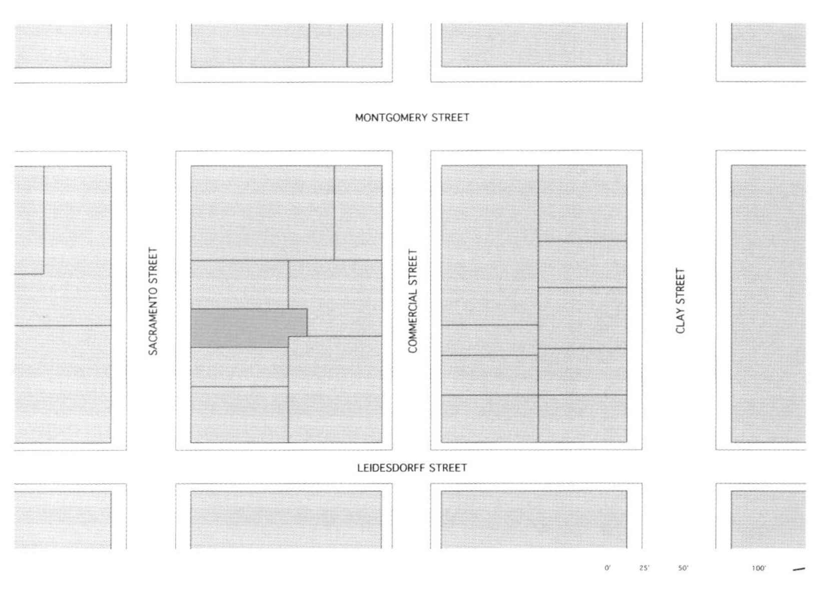

SITE PLAN-SAN FRANCISCO FINANCIAL DISTRICT

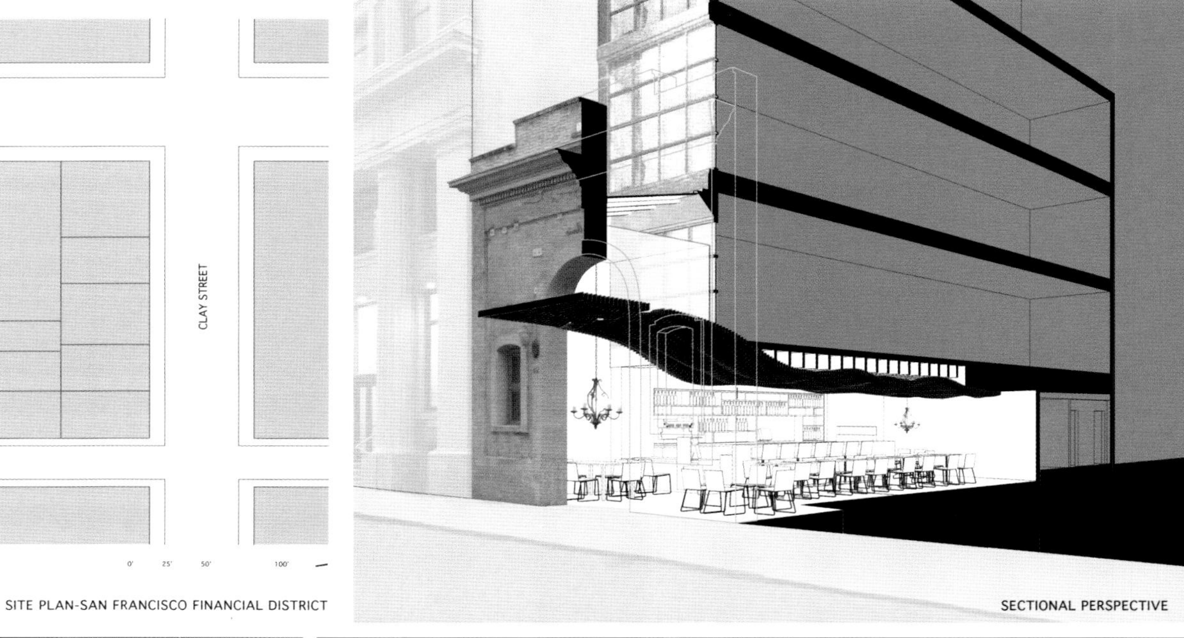

SECTIONAL PERSPECTIVE

INTERIOR CONCEPT

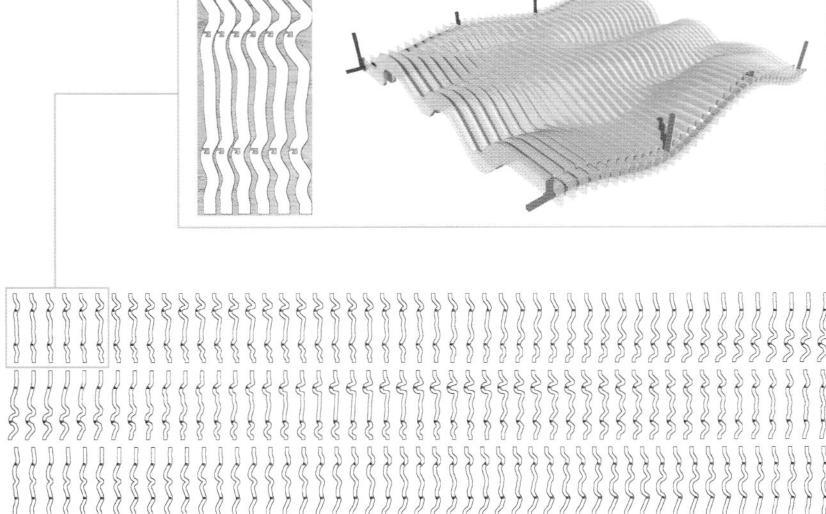

CANOPY-3D MODELING AND LASERCUT FABRICATION

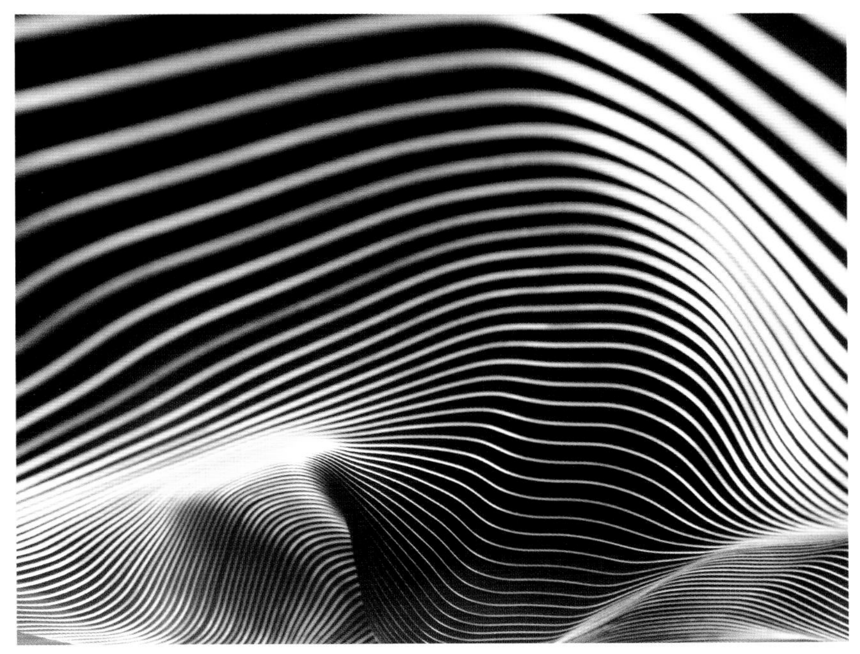

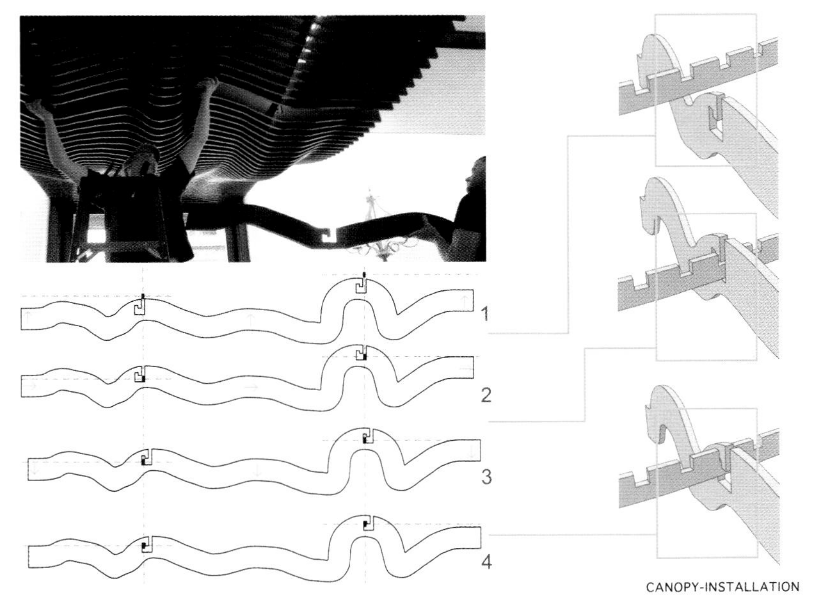

CANOPY-INSTALLATION

Aidlin Darling Design公司负责设计了Wexler's 餐厅的内部装修，色调以保守的辛白色和斑斑驳驳的橡树色为主，间接照明营造出恬淡、温暖的感觉。

古朴的选材赋予餐厅浓郁的乡土气息，而这又恰恰衬托出菜品的特色。淡淡的烟，微醺的木，由激光切割的MDF散热片构成的、呈波状起伏的天花板，动感十足，恰到好处地完成了从“古朴”到“现代”的过渡。

该设计方案包括四个基本要素，共同营造了温暖、明亮的餐厅氛围，使菜品更加出色，同时也为传统的烧烤方式注入了创新精神。

内部装饰色调以不张扬的橡树色和辛白色为主，彰显着回归田园的风格。对这些传统耐用材料的低调处理办法让人不由自主地想起了传统的烧烤。沿袭相同的思路，设计师大胆地使用了深红色，把传统的铁艺枝形吊灯做成极具现代气息的灯具。温馨的照明和独特的建筑符号相得益彰，将田园风格演绎为一个私密的空间，在那里，客人可以尽享无可挑剔的美食。

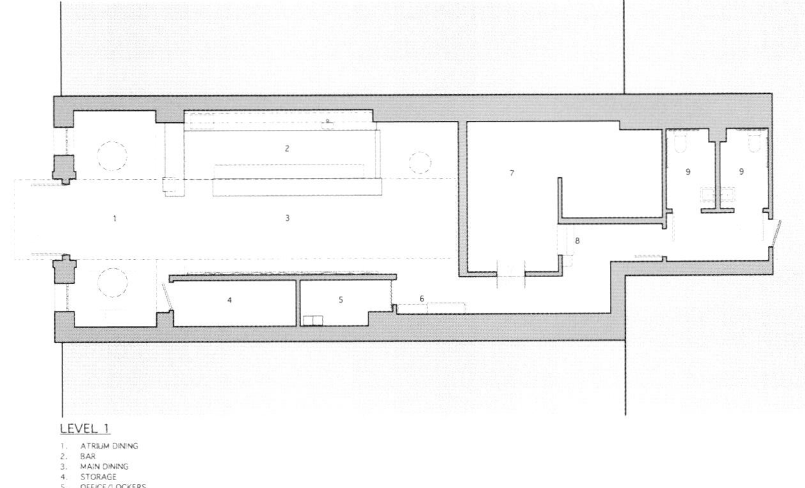

FLOOR PLAN

KAZUMI SUSHI LOUNGE

Designers: Jean de Lessard, Stéphane Proulx
Design Company: Jean de Lessard, Designer Créatif
General Contractor: Rénovatech. M. Nicolas Khuu
Design Consultant: Kitchen Equipement JF. Jacques Brazeau
Location: Montreal, Canada
Area: 320 m^2
Photographer: Jean Malek

Origami is the ancient and celebrated Japanese art of paper folding: by folding one or several sheets of paper, the designer creates a shape, a decoration. This art originated in ceremonies in which folded pieces of paper were used to ornament jugs of sake. It is in this spirit of art and celebration that the Jean de Lessard design firm approached the interior design of the Kazumi Japanese restaurant.
The designer's intuition was perfect: the angular shape, pierced with sushi-pink translucent glass on its own, creates the ambience and tone of the restaurant. The colorful opening also gives a glimpse of the intense work being done in the kitchen.
A colorful and fresh locale – just like Chef Tri's cuisine!
As soon as they enter, guests are enveloped in a corridor of clear glass set into brightly striped walls. The vinyl covering, in a motif created by the designer and factory-printed, invariably draws the eye toward the sushi counter, the restaurant's main room.
A challenging layout
Dealing with a rectangular space without a large opening to the exterior, Jean de Lessard had to use imagination to work with angles of view and create dynamic visual points of impact, while maintaining an overall impression of simplicity and balance. By using dark curtains to dampen the sound level and playing with the transparency of certain walls, the designer give some materials a double function, thus optimizing his design.
The essence of Kazumi
With a contemporary approach and through use of color, Jean de Lessard transposed the essence of Restaurant Kazumi into its interior design: a colorful cuisine in a stimulating, friendly setting.

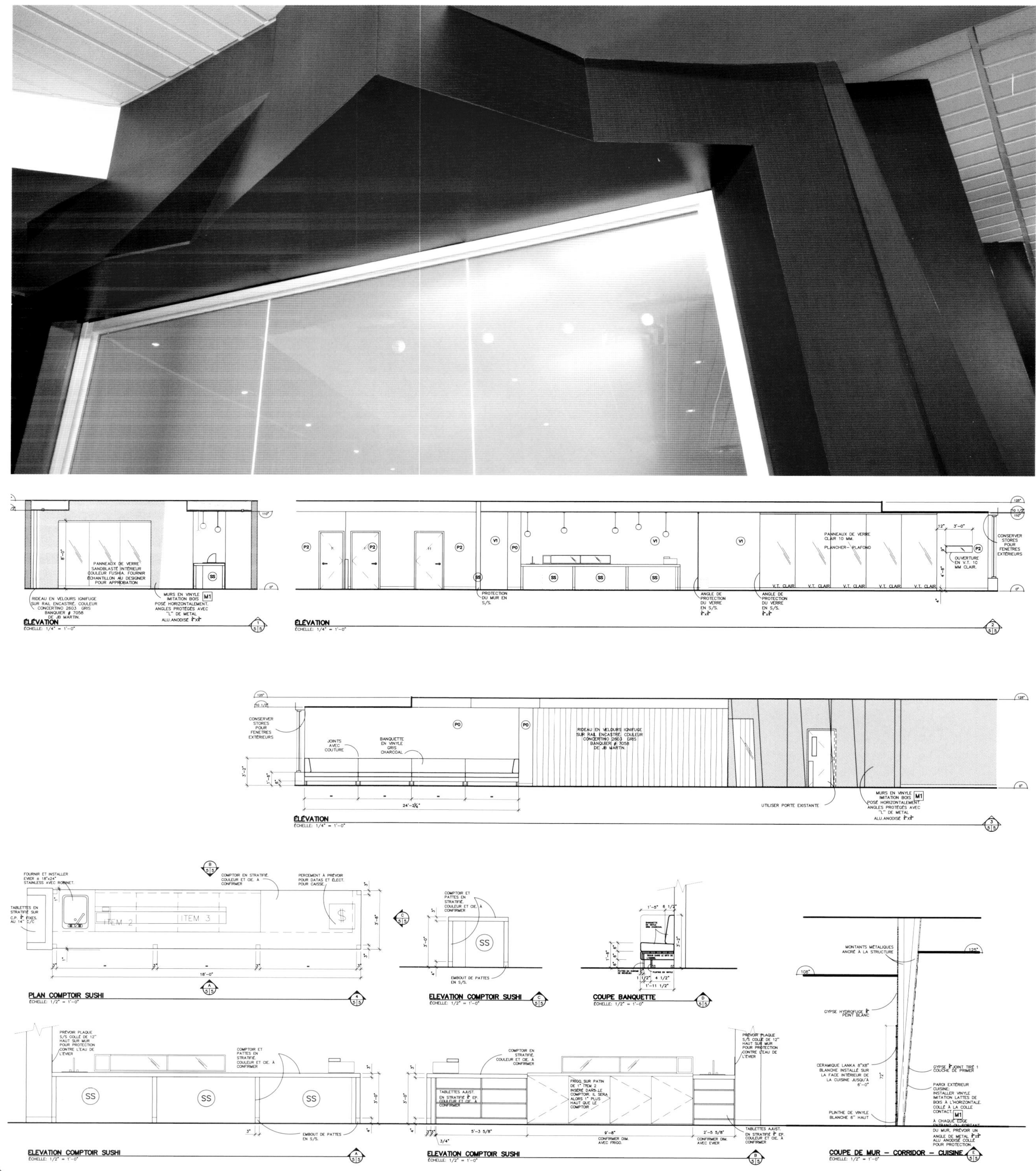

PANNEAUX DE VERRE SANDBLASTÉ INTÉRIEUR COULEUR FUSHIA. FOURNIR ÉCHANTILLON AU DESIGNER POUR APPROBATION
8'-0"
SS
110"
0"
RIDEAU EN VELOURS IGNIFUGE SUR RAIL ENCASTRÉ. COULEUR CONCERTINO 2603 GRIS BANQUIER # 7058 DE JB MARTIN.
MURS EN VINYLE IMITATION BOIS M1 POSÉ HORIZONTALEMENT. ANGLES PROTÉGÉS AVEC "L" DE METAL ALU.ANODISÉ ¾"x¾"
ÉLEVATION
ÉCHELLE: 1/4" = 1'-0"
P2
V1
P0
PROTECTION DU MUR EN S/S.
ANGLE DE PROTECTION DU VERRE EN S/S. ¾"x¾"
PANNEAUX DE VERRE CLAIR 10 MM.
PLANCHER- PLAFOND
V.T. CLAIR
12"
3'-0"
OUVERTURE EN V.T. 10 MM CLAIR.
CONSERVER STORES POUR FENETRES EXTÉRIEURS
125"
10 1/2"
4'-6"
ÉLEVATION
ÉCHELLE: 1/4" = 1'-0"
CONSERVER STORES POUR FENETRES EXTÉRIEURS
JOINTS AVEC COUTURE
BANQUETTE EN VINYLE GRIS CHARCOAL
RIDEAU EN VELOURS IGNIFUGE SUR RAIL ENCASTRÉ. COULEUR CONCERTINO 2603 GRIS BANQUIER # 7058 DE JB MARTIN.
3'-2"
1'-6"
6"
24'-2¼"
UTILISER PORTE EXISTANTE
MURS EN VINYLE IMITATION BOIS M1 POSÉ HORIZONTALEMENT. ANGLES PROTÉGÉS AVEC "L" DE METAL ALU.ANODISÉ ¾"x¾"
ÉLEVATION
ÉCHELLE: 1/4" = 1'-0"
FOURNIR ET INSTALLER EVIER ± 18"x24" STAINLESS AVEC ROBINET.
TABLETTES EN STRATIFIÉ SUR C.P. ¾" FIXES. AU 14" C/C
COMPTOIR EN STRATIFIÉ. COULEUR ET CIE. À CONFIRMER
PERCEMENT À PRÉVOIR POUR DATAS ET ÉLECT. POUR CAISSE.
ITEM 2
ITEM 3
$
3'-6"
18'-0"
PLAN COMPTOIR SUSHI
ÉCHELLE: 1/2" = 1'-0"
COMPTOIR ET PATTES EN STRATIFIÉ. COULEUR ET CIE. À CONFIRMER
SS
3'-0"
EMBOUT DE PATTES EN S/S.
ELEVATION COMPTOIR SUSHI
ÉCHELLE: 1/2" = 1'-0"
1'-5"
6 1/2"
3'-2"
1 1/2"
4 1/2"
1'-11 1/2"
COUPE BANQUETTE
ÉCHELLE: 1/2" = 1'-0"
PRÉVOIR PLAQUE S/S COLLÉ DE 12" HAUT SUR MUR POUR PROTECTION CONTRE L'EAU DE L'ÉVIER
COMPTOIR ET PATTES EN STRATIFIÉ. COULEUR ET CIE. À CONFIRMER
SS
EMBOUT DE PATTES EN S/S.
ELEVATION COMPTOIR SUSHI
ÉCHELLE: 1/2" = 1'-0"
COMPTOIR EN STRATIFIÉ. COULEUR ET CIE. À CONFIRMER
TABLETTES AJUST. EN STRATIFIÉ ¾" EP. COULEUR ET CIE. À CONFIRMER
FRIGO SUR PATIN DE 1" ITEM 2 INSÉRÉ DANS LE COMPTOIR. IL SERA ALORS 1" PLUS HAUT QUE LE COMPTOIR
PRÉVOIR PLAQUE S/S COLLÉ DE 12" HAUT SUR MUR POUR PROTECTION CONTRE L'EAU DE L'ÉVIER
5'-3 5/8"
3/4"
9'-8"
CONFIRMER DIM. AVEC FRIGO.
2'-5 5/8"
CONFIRMER DIM. AVEC EVIER
TABLETTES AJUST. EN STRATIFIÉ ¾" EP. COULEUR ET CIE. À CONFIRMER
ELEVATION COMPTOIR SUSHI
ÉCHELLE: 1/2" = 1'-0"
MONTANTS MÉTALIQUES ANCRÉ À LA STRUCTURE
125"
108"
GYPSE HYDROFUGE ⅝" PEINT BLANC
CERAMIQUE LANKA 8"X8" BLANCHE INSTALLÉ SUR LA FACE INTÉRIEUR DE LA CUISINE JUSQU'À 6'-0"
72"
PLINTHE DE VINYLE BLANCHE 6" HAUT
GYPSE ⅝" JOINT TIRÉ 1 COUCHE DE PRIMER
PAROI EXTÉRIEUR CUISINE: INSTALLER VINYLE IMITATION LATTES DE BOIS À L'HORIZONTALE. COLLÉ À LA COLLE CONTACT. M1
À CHAQUE EDGE ENTRANT OU SORTANT DU MUR, PRÉVOIR UN ANGLE DE METAL ¾"x¾" ALU ANODISÉ COLLÉ POUR PROTECTION.
COUPE DE MUR – CORRIDOR – CUISINE
ÉCHELLE: 1/2" = 1'-0"

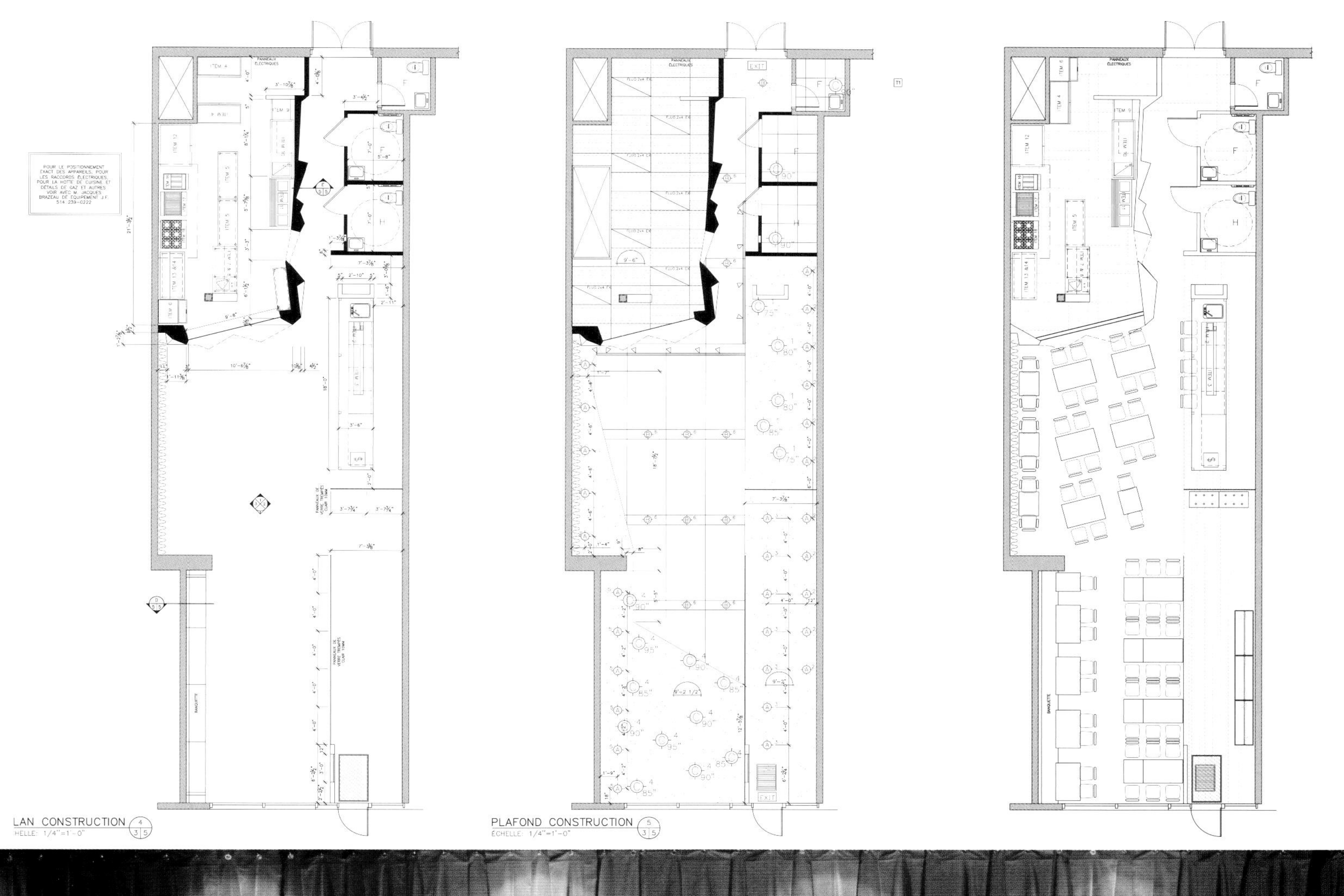
LAN CONSTRUCTION
HELLE: 1/4"=1'-0"
PLAFOND CONSTRUCTION
ECHELLE: 1/4"=1'-0"

Origami是日本古老而闻名的折纸艺术：把一张或者几张纸进行折叠，创造出一种形状，用于装饰，这种折纸艺术最早用于庆典时装饰清酒瓶。正是基于这种艺术灵感，Jean de Lessard设计公司对Kazumi日式餐厅进行了内部设计。

设计师的直觉是完美的：棱角分明的形状搭配类似寿司的粉色半透明玻璃，营造出了餐厅的气氛和基调。彩色的入口也能让客人看到厨房里正在进行的紧张工作。

多彩、清新的场所——就像是主厨Tri的菜肴一样！

客人一进来，就像是置身于一个明亮的水晶宫一样，四周玻璃墙的条纹装饰清晰明快。乙烯基的覆盖物，是设计师设计的图案，由工厂印制完成，总是吸引着客人走向餐厅的主要空间——寿司吧台。

挑战性的布局

在没有宽阔的入口门厅情况下，设计一个长方形的空间，Jean de Lessard的设计公司不得不发挥想象力，不仅要打造出动态的视觉冲击效果，而且要保持餐厅整体的简洁和平衡。使用深色的窗帘来缓冲声音大小，运用透明的墙体，设计师赋予这些材料双重功效，优化了设计。

Kazumi餐厅的理念

Jean de Lessard设计公司使用现代手法，辅以色彩的应用，成功地将Kazumi餐厅的理念转化成内部设计：在新奇、友好的氛围中享用色香味俱全的菜肴。

SPICE MARKET RESTAURANT

Designers: Rob Wagemans, Jeroen Vester, Melanie Knüwer, Ulrike Lehner, Erik van Dillen, Sofie Ruytenberg, Femke Zumbrink
Developer: McAleer & Rushe Group
Area: 530 m^2
Photographer: Ewout Huibers
Clients: Starwood Hotels & Resorts, McAleer & Rushe Group, Culinary Concepts by Jean-Georges

Imagine a wall of spices containing all the colours, flavours and fragrances of Asian cuisine. Imagine a spice cabinet, two floors high, twenty-four metres long and revealing every ingredient the chef will need to create the distinguished Spice Market dishes. This spice cabinet is here, as the center-piece of the restaurant and starting point for a great dinnertime. When looking up from the street and see through the transparent restaurant façade, guests can even see the cabinet from there.

Spice Market London is a unique mix of the ethnic vintage feel of Spice Market New York and the contemporary architecture of the new building on Leicester Square in London. The eclectic and intimate design is a result of gold mesh sliding screens, brass screen lanterns, jatoba timbo flooring and cosy booths, a unique brass "birdcage"spiral stairs and 600 wok-lights.

The restaurant is a two-level space connected by the grand iconic birdcage-staircase and a central void in the mezzanine floor, transporting the energy from the kitchen into the entire space. Guests can settle on both floors and choose from a wide variation of seats: cocktail bar, sushi bar and lounge seating on the ground floor and restaurant seating with an open kitchen on the mezzanine floor.

Monumental brass lanterns with laser-cut patterns give a warm and soft glow and shed an extraordinary decorative light throughout the restaurant area. Another eccentric light creation can be found when looking up into an endless sea of wok lamps hanging from the ceiling. Of course, Spice Market is not just about the cooking inside the kitchen.

There are bar stools if guests have to wait for a friend and have a drink in the meantime. There are armchairs at the bespoke dinner table for two, a black leather seating booth for the more intimate dinners and a black leather lounge chair or sofa with cushions to relax and have a chat over the coffee later.

想象一下，一面墙上摆满了颜色各异、口味齐全、香气四溢的亚洲美食香料！还有一个香料橱柜，高达两个楼层，长24米，陈设着厨师烹调Spice Market菜品所需的各种料汁。调料柜是餐厅的核心，也是享受美食的关键所在。临街仰望，透视上下贯通的餐厅店面，甚至可以看见这个调料柜。

Spice Market伦敦店结合了传统的纽约店的风格和伦敦现代派的莱斯特广场建筑风格于一体。金色网格的滑动屏幕、铜色屏幕灯、兰叶苏木藤条地板、舒适的摊亭、一段铜色的鸟笼状螺旋式楼梯和600个WOK灯构成了这个折中又私密的空间。

这个餐厅是一个两层建筑，由一个巨大的鸟笼式楼梯和楼梯间的夹层空间连接起来，把厨房里火热的气氛传递到餐厅的各个空间。客人可以在两层餐区自由选择座位：鸡尾酒吧区、寿司区、一楼大厅或是位于夹层的敞开式厨房区。

标志性的铜色灯具，灯具上激光切割的条纹散发着温暖柔和的光，映射出一个贯通上下的华美灯饰。抬眼望向繁星般的WOK灯区，另一个不同寻常的灯饰自天花板悬垂而下。总之，在Spice Market中，这个敞开式厨房是夺人眼目的一大亮点。

如果是等朋友，可以在酒吧区坐在高脚凳上先喝上一杯；双人预订就餐区还有扶手椅，黑色皮座的包间适合亲密朋友的约会。黑色皮座的长椅或摆放着靠垫的沙发让人不禁想要手捧咖啡，惬意享受。

NICOLA RESTAURANT

Designer: Marcos Samaniego Reimúndez
Design Company: Mas • Arquitectura
Photographer: Ana Samaniego

Italinan flavour and "al dente"shape.
Good food and careful design are responsible of this Italian restaurant's success.
Nowadays, serve good food is not a guarantee of success. People are looking for special spaces where enjoy their free time. For this reason, restaurant´s design has become essential. Success is at stake. Marcos Samaniego, the main architect from Mas Arquitectura, bets in this kind of project for developing functional spaces but with powerful brand image.
This restaurant, placed on A Coruña (Spain), follows two basics premises: intimacy for diners and clear organization for workers. The local is distributed around axes of movement distinguished by furniture´s distribution. As a result, "Nicola restaurant", provides best facilities to work and, at the same time, is perceived by customer as a tidy and clean space.
The entrance offers modern image based on simple lines. Inside, the main space is used as dining room. Different materials, as wood, paper vinyl, tile and silestone are used in order to obtain a balance result of quality, durability and warmth. Indeed, the restaurant has got an extra space ideal for couples.
"Nicola restaurant" is also distinguished for its personality. Different spaces are designed reflecting different guests' preferences. So, after seeking comfort in central space and privacy couples´ area, designers need to create a space for calm people. This private room has been decorated with a collage which includes famous Italian monuments and other from European countries.
Mas Arquitectura not only has developed interior design and execution but also has created the brand image. As a result, all elements are integrated in one atmosphere.

这是一个意大利口味的、外形是“弹牙”状的餐厅。

精美的食物和精心的设计相得益彰，共同成就了这家意大利餐厅。

现在，餐厅仅仅能够提供美食已经不能立足市场。人们追求的是能够享受闲暇时光的个性场所。为此，餐厅的设计尤为重要。设计师在保持原有强大的品牌形象的同时，冒险突出了餐厅的功能。

坐落于A Coruña的Nicola餐厅，在设计时遵循两个原则：满足客人对私密性的要求和工作的透明性，内饰格局参照家具分布并以其为中心。Nicola餐厅为员工提供最好的工作设施，同时其整洁、干净的就餐环境也得到客人的赞许。

入口依托简洁的条纹显得非常摩登。内部主区用作就餐大厅。不同的材料，例如，实木、纸质乙烯树脂，以及瓷砖的使用保证了餐厅的质量和耐用性及保温性。事实上，餐厅还另设双人雅间。

Nicola餐厅也因其人性化的设计而与众不同。不同的就餐区满足了不同客人的需求。所以，为了能使喜欢清静的客人在市区中心享受到一个私密的空间，设计师特意打造了这样一个空间。这个私密空间的内部装饰是意大利和其他欧洲国家的著名的拼贴画。

设计师不但完成了室内设计，而且还扩大了其品牌的影响力，因此，所有元素彼此共融，共同营造了和谐的氛围。

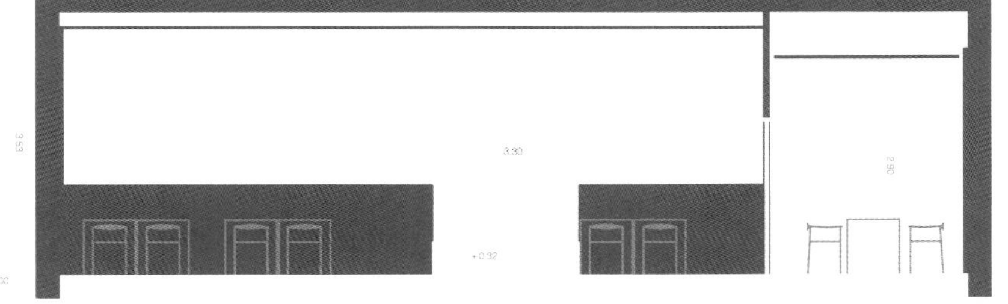

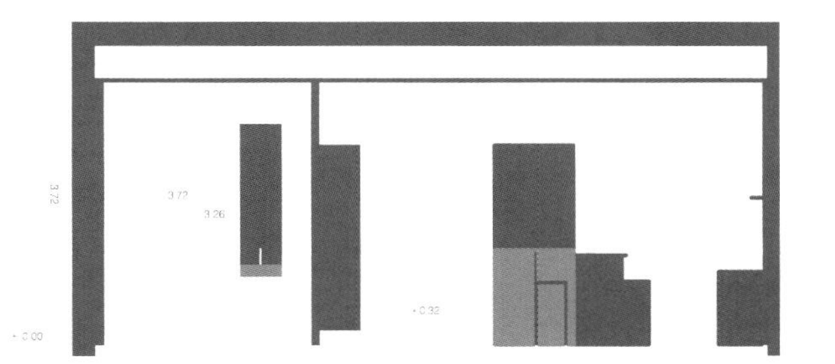

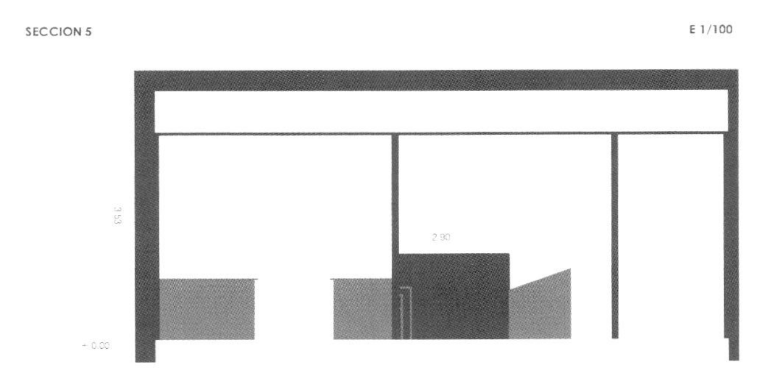

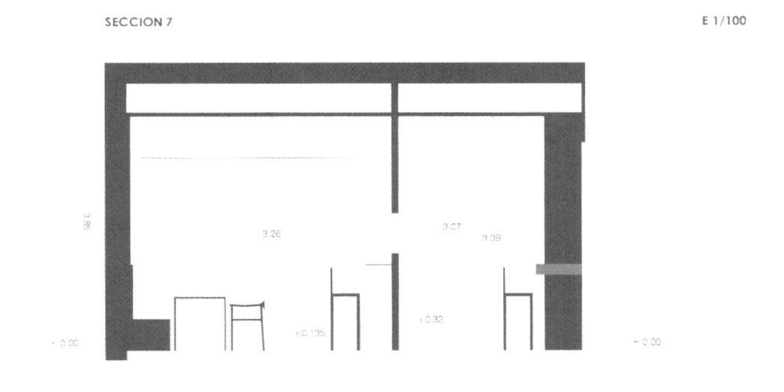

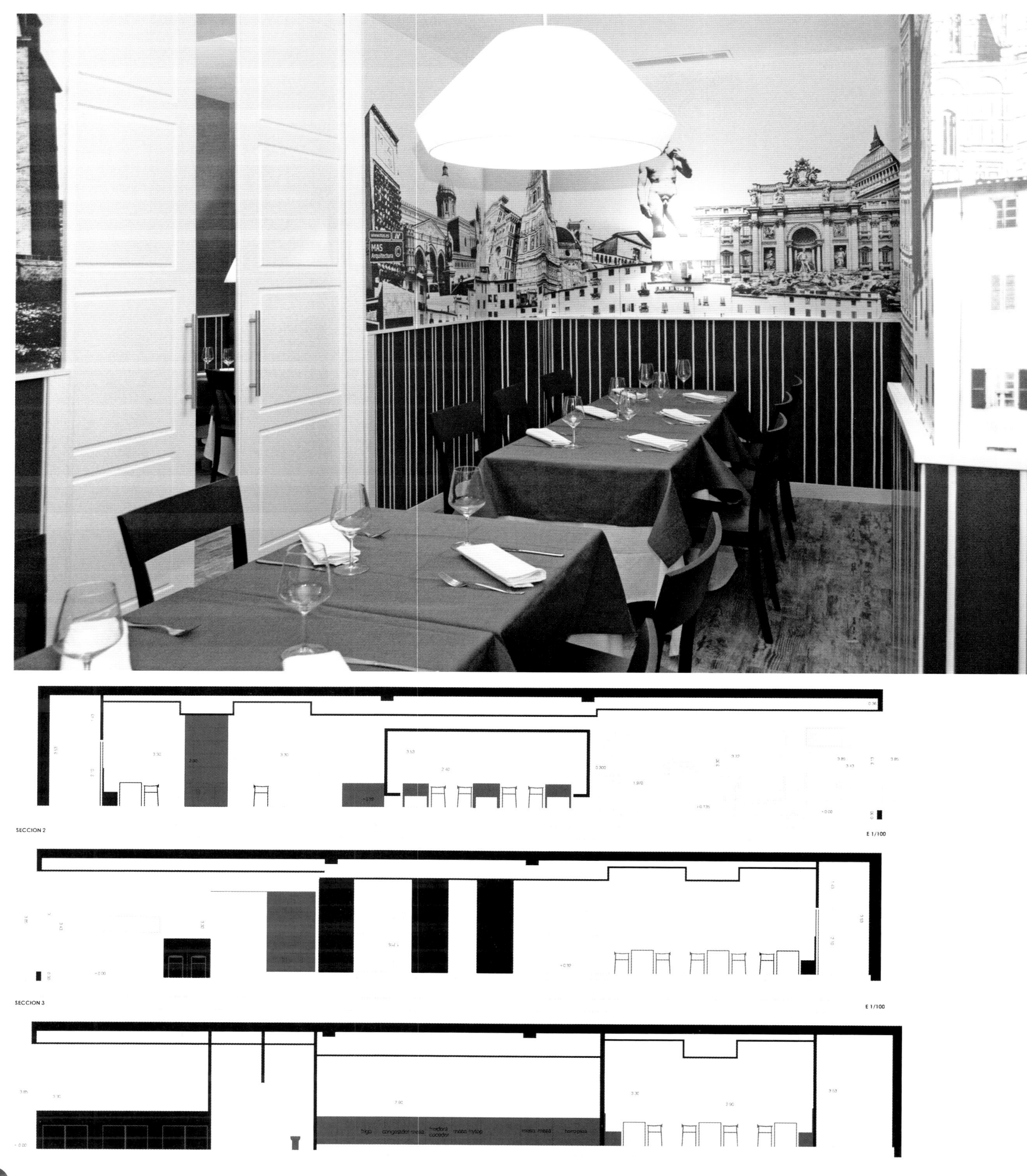
MAS
Arquitectura
SECCION 2
E 1/100
SECCION 3
E 1/100
congelador mesa
cocedor

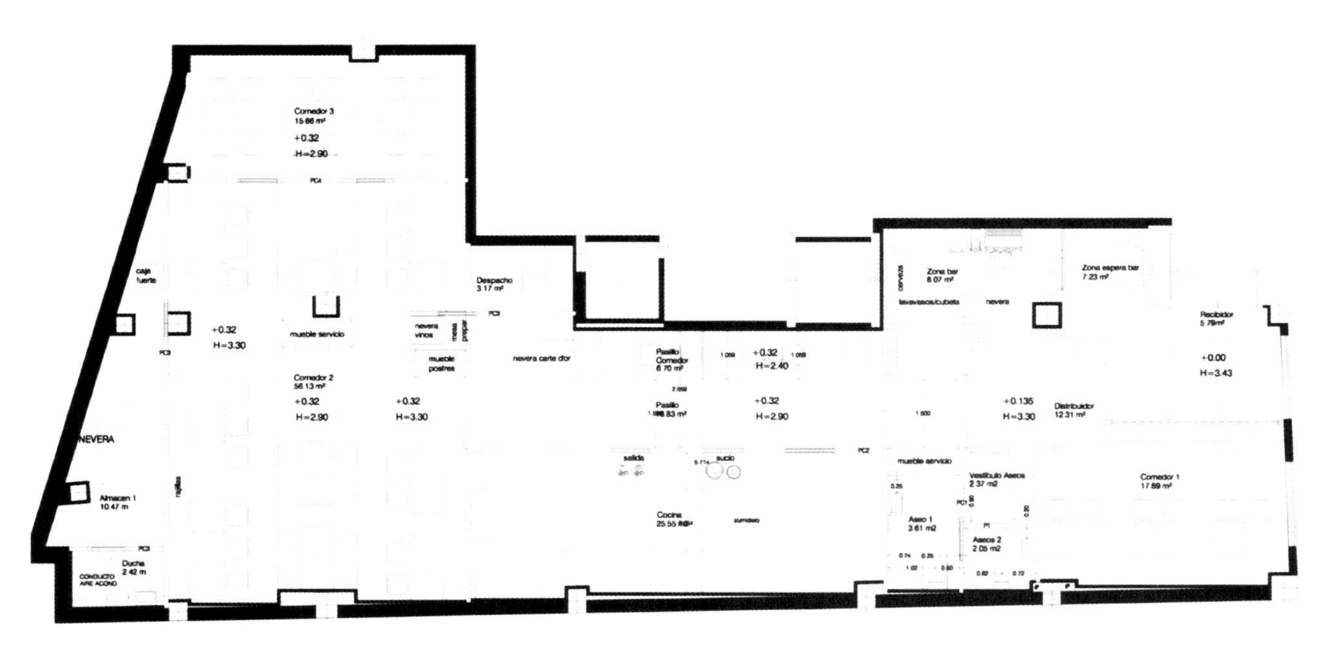
Comedor 3
15.66 m²
+0.32
H=2.90
caja fuerte
Despacho
3.17 m²
+0.32
H=3.30
mueble servicio
nevera vinos
mesa prepar
mueble postres
nevera carte d'or
Pasillo Comedor
6.70 m²
+0.32
H=2.40
Comedor 2
56.13 m²
+0.32
H=2.90
+0.32
H=3.30
Pasillo
+0.32
H=2.90
NEVERA
salida
sucio
Almacen 1
10.47 m
Cocina
sumidero
Ducha
2.42 m
CONDUCTO AIRE ACOND.
cerveza
Zona bar
8.07 m²
Zona espera bar
7.23 m²
lavavasos/cubeta
nevera
Recibidor
5.79m²
+0.00
H=3.43
+0.135
H=3.30
Distribuidor
12.31 m²
mueble servicio
Vestíbulo Aseos
2.37 m2
Comedor 1
17.89 m²
Aseo 1
3.61 m2
Aseos 2
2.05 m2

HOLYFIELDS

Designer: Ippolito Fleitz Group - Identity Architects
Design Company: Ippolito Fleitz Group - Identity Architects
Design Team: Gunter Fleitz, Peter Ippolito, Michael, Bertram, Bartlomiej, Pluskota, Tilla Goldberg, Tim, Lessmann, Moritz Koehler, Joerg Schmitt, Joss Haenisch
Lighting Design: Pfarré Lighting Design, Munich
Location: Frankfurt, Germany
Area: 459 m^2
Photographer: Zooey Braun

Holyfields, a wholly new restaurant chain concept, commissioned our studio to develop a modular, scalable space system with a distinctive look and feel. With its innovative food concept, Holyfields sets new accents in system gastronomy for a discerning, urban clientele. It satisfies the need for fast and delicious, great value food, as well as providing a visual and atmospheric dining highlight. The differentiated range of seating zones caters towards the different needs of its guests, who are certain to find the right setting for anything from a quick snack to a meal out with friends or a bigger celebration.

PETER
PAUL
MARY
holyfields
holyfields

time to eat

Holyfields是一个具备全新餐饮理念的连锁餐厅。业主委托设计师为其设计独特的、模块化的形象与空间。凭借其创新的餐饮理念，Holyfields为挑剔的业主带来全新的美食体验。这里的食物满足快捷、美味、营养的需求，同时，视觉冲击力也足够强。不同的座位区满足了各种客人的需求，不论是出来吃个便餐、朋友小聚，又或是进行更大的聚餐。

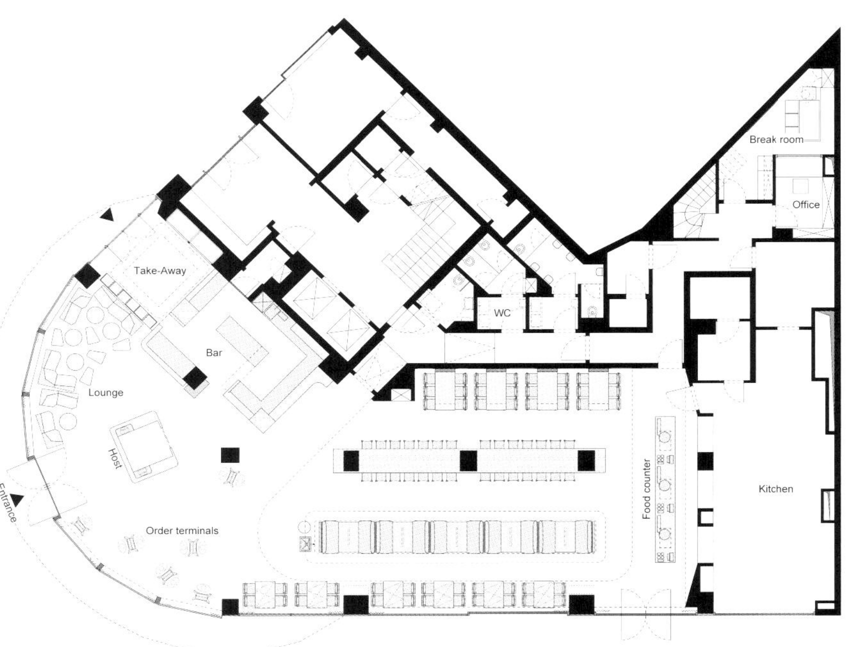

KITH CAFE

Designer: Hjgher
Design Company: Hjgher
Location: 7 Rodyk Street, #01-33, Singapore
Area: 28 m^2
Photographer: the Shooting Gallery

The HJGHER's concept for Kith Cafe is inspired by a simple phrase, taken off the Japanese film "Be with you". It translates into "would you care to have coffee with me sometime?", a plain and sincere question that has the power to turn strangers into kith and kin. It was not just a matter of coming up with a Kith Cafe, but creating a complete experience. This meant no fancy lampshades, mass-manufactured furniture, or unnecessarily printed materials were to be used in its interior. Instead, the designers focused on delivering an unassuming space that could support Kith's honest philosophy and ethos.

The custom-made furniture for the seating area was designed to be flexible and movable, to break down personal space barriers and encourage customers to interact. To maximize floor space, most modular blocks can double as either a table or a bench, and even used to seat two people instead of one as needed. This way, the small space can accommodate a group of up to twenty.

Emphasis was also placed on relaying the sincerity and openness of the cafe: open wrought iron shelves openly display the storage of fresh produce and most of the outlet's inventory. In line with the informal atmosphere, the space is adorned by one sole piece of "decoration" – a handwritten menu on the chalkboard wall, where old specials are casually wiped off with new ones scribbled on daily.

新加坡Kith咖啡馆的室内设计灵感来自于日本电影《与你同在》，意为“你有空跟我喝杯咖啡吗？”。朴素、真诚的话语拉近了陌生人之间的距离，从此便相识相知。它是一个能够令人触景生情的咖啡馆。这里没有喧宾夺主的台灯、整齐划一的设备和花哨的装饰。相反，设计师营造出来一个内敛而本位的空间。

座位区定制的桌椅活动自如，打破障碍，客人可以自由交流，为了使空间最大化，多数大块模板可以重叠摆放，兼具桌凳功能，甚至可以充当双人座。这样，这个不大的空间的客容量就高达20人。

咖啡馆的诚信经营和透明化也得以强调。敞开式铁架上摆放着现场磨出的咖啡和本店的独家新品。与之相呼应的是，兼具黑板功能的墙上是每天手写的菜单。

YUKON BAY IN ADVENTURE ZOO HANOVER

Designer: Dan Pearlman Erlebnisarchitektur GmbH
Location: Erlebnis-Zoo Hannover, Adenauerallee 3, 30175 Hannover, Germany
Area: Yukon Bay: 26, 500 m²
Shop within Yukon Bay: 184 m²

Yukon Bay, opened last May in the German Hanover Zoo, represents 25,000 m² of Canadian landscape and new homes for over 100 animals and 15 species.
Apart from the animal attractions, the gastronomic center of Yukon Bay, the Market Hall, offers a unique one of the visitors main attractions.
The red timber hall is based on the typical Yukon harbor architecture, intuitively telling the story of an old 1920s fish warehouse, converted sometime in the 1950s into a market hall with stands and storage facilities, decorated with swordfish and other authentic fishing memorabilia in the tradition of the harbor town of Yukon Bay. Here, the Klondike Grill's Giant Bison Burger became legendary, along with Canadian inspired recipes such as crusty spare ribs, special steaks and golden chicken wings. The restaurant offers seating for 480 guests, and a further 420 seats are available outside.
For those who are looking for fresh Alaska fish, Mike's Fish Inn is the perfect choice. According to the story, Mike, the son of German immigrants, bought the old fish market and built a new gallery – the Mike's Fish Inn.
All over the fish market hall, the traces of the past are plain to see: faded notices of fish deliveries on the walls, partitions between the old trader's booths, and even a drain cover with its cast inscription recalls the former use of this imposing building. Here, the smoked fish – straight from the smokers – and the fish soup are the specialties.
For snacks,there's no need to go too far. The Corner Sandwich creates fresh baguettes, bread, oats, sesame flat bread and bagels that continue the Yukon Bay experience.
The most difficult is to choose the fillings: smoked trout fillets, smoked salmon, hot smoked shrimp or salmon, smoked herring, pickled herring, salmon or butter fish, grilled chicken breast, elk salami, honey ham, cranberry-cheese, turkey breast, smoked pork or bison ham. All of this is always rounded off with fresh lettuce and crunchy cheese primer.
For the sweet tooth, it's better to go to the Kathy's Cake. This stand specializes in the cozy moments, typical of an American coffee shop.
But the temptations are not over. Last, the gelateria. The famous Luigi Amarone Italian homemade ice creams are one of the specialties of Yukon Bay.

Best homemade Ice Cream

Luigi Amarone
GELATERIA · CAFÉ · BAR

Best homemade Ice Cream

Vespa
Amarone

YUKON
YUKON
SANDWICH
YUKON MARKET HALL

BISON
KLONDIKE GRILL
Spare Ribs, Wings & Co.
BURGER
BEST OF POTATOE
EXTRAS
BEST OF POTATOE
SPARE RIBS & CO

SANDWICH, BAGEL & CO
FISH AND FRIENDS
SANDWICH CORNER
FRESH & TASTY
FARMERS MARKET
BAKED

FREEZER 2
YUKON
YUKON

Yukon Bay位于加拿大的德国汉诺威动物主题公园内，占地2.5万平方米，是100多只动物和15个物种的新家园。

除了有动物吸引游客之外，Yukon Bay的烹饪中心、贸易大厅也以其别具一格的魅力，吸引着八方来客。

红色漆木大厅建在独特的Yukon Bay上，厅内饰以旗鱼和逼真的Yukon Bay城传统装饰物，讲述19世纪20年代它作为储鱼仓库的历史，在20世纪50年代增设摊位和存储设施，演变为交易大厅的故事。在这里，吸取了加拿大配方Klondike Grill的巨型野牛汉堡透漏着一股传奇色彩，例如，脆皮肋排、特色牛排、黄金炸鸡翅等。店内有480个席位，另外，室外还可容纳420人就餐。

Mike’s Fish旅店是那些爱吃阿拉斯加鱼的人的不二选择。故事是这样说的:麦克，一个德国移民的后裔，买下了老旧的鱼市，还新建一个大厅，这就是现在的Mike’s Fish旅店。

大厅在诉说着岁月的痕迹：墙上张贴卖鱼告示，老旧的摊厅间的隔墙，刻有出海祭词的排水井盖，这一切都在告诉世人它辉煌的历史。

品尝小吃无需走得太远。Corner Sandwich沿袭Yukon Bay风格，自制法棍、面包、燕麦粥、西麦饼和硬面包圈。

最难选的还是主食：熏鳟鱼片、熏大马哈鱼、香辣熏虾、香辣大马哈鱼、熏鲱鱼、腌制鲱鱼、鲜大马哈鱼、鲜鲳鱼、烤鸡胸脯肉、意大利麋鹿肠、蜂蜜火腿、越橘芝士、火鸡胸脯肉、熏猪排、野牛火腿。所有的肉都可以用新鲜的生菜叶夹裹着来吃，再配上鲜脆的芝士，美味无穷。

喜爱甜食的人们，请去Kathy’s Cake一饱口福。这家小店精于营造经典的美式休闲氛围，舒适又惬意。

但是诱惑还远远没有结束，最后一站是gelateria，这里有意大利著名的路易吉•阿玛尼亚自制冰激凌，是Yukon Bay的一大特色。

WC

Pasta
Sweets

Ob alt ob jung, ob
groß ob klein,
Pasta schmeckt doch jedem
fein.

NAUTILUS PROJECT

Designer: Yuhkichi Kawai
Design Company: Design Spirits Co. , Ltd.
Location: ION Shopping Mall, Singapore
Area: 502.75 m^2
Lighting Consultant: Muse-D Inc. Kazuhiko Suzuki, Misuzu Yagi
Photographer: Toshihide Kajiwara

The Nautilus Project is located on the fourth floor of the ION shopping center, where opened recently on the Orchard Road, Singapore. This floor is not only the restaurant floor, but also shops as well. This floor is a composite both sales and drinks and eats. The beginning of the plan, the food consultant prepared a certain concept, chef, designer, and location, so the designer had a difficulty in leading my ideal interior design.
The owner is the president of the cargo company and surprisingly a beautiful woman boss, so the designer decided to reflect her sophisticate, elegant, and tender characteristic to this restaurant project. The chef was done recruiting of by owner oneself from New Zealand, and the chef was popular among New Zealander like a celebrity.
The designer predicted that visitor unit price would rise in meetings, so he stopped making a façade but designed the entrance where a common use passage led to restaurant to make visitor enter easily. The skeleton of the restaurant had curve, so he made the others curve to let it be familiar. The wood material was requested by owner. The designer made effort to let it not look futuristically because of its curve. It's a design,not only being on half of the designer's creation,but also standing the rest of time.

the nautilus project

the nautilus project

Nautilus餐厅位于新加坡Orchard路的ION购物中心四楼。这一层不仅有餐厅，也有购物商场。在这里，客人既可以享受美食也可以畅快购物。在该项目实施之初，餐厅顾问就已经对于厨师、设计师、餐厅位置的选择有了一些自己的想法。所以，该餐厅在内部装饰方面也颇费心思。

该餐厅的业主是货运公司总裁，而且是一位美丽的职业女性。所以，设计师想把她的成熟老练、大方优雅，以及细腻的情感特质全都通过该餐厅的设计反映出来。厨师是老板娘亲自从新西兰聘请回来的，他在新西兰的饮食界颇为出名。

入口处特别设计了一个长廊，非常便于客人在就餐高峰期的有序进入。该餐厅的原址本来就有回廊，所以，设计师在别处也设计了相似的回廊。业主希望使用木质材料来装饰餐厅，因此，设计师做了最大努力，从而使该餐厅看上去不至于太新潮。它不仅可以代表设计师的创意，还能够经得起时间的考验。

EXIT

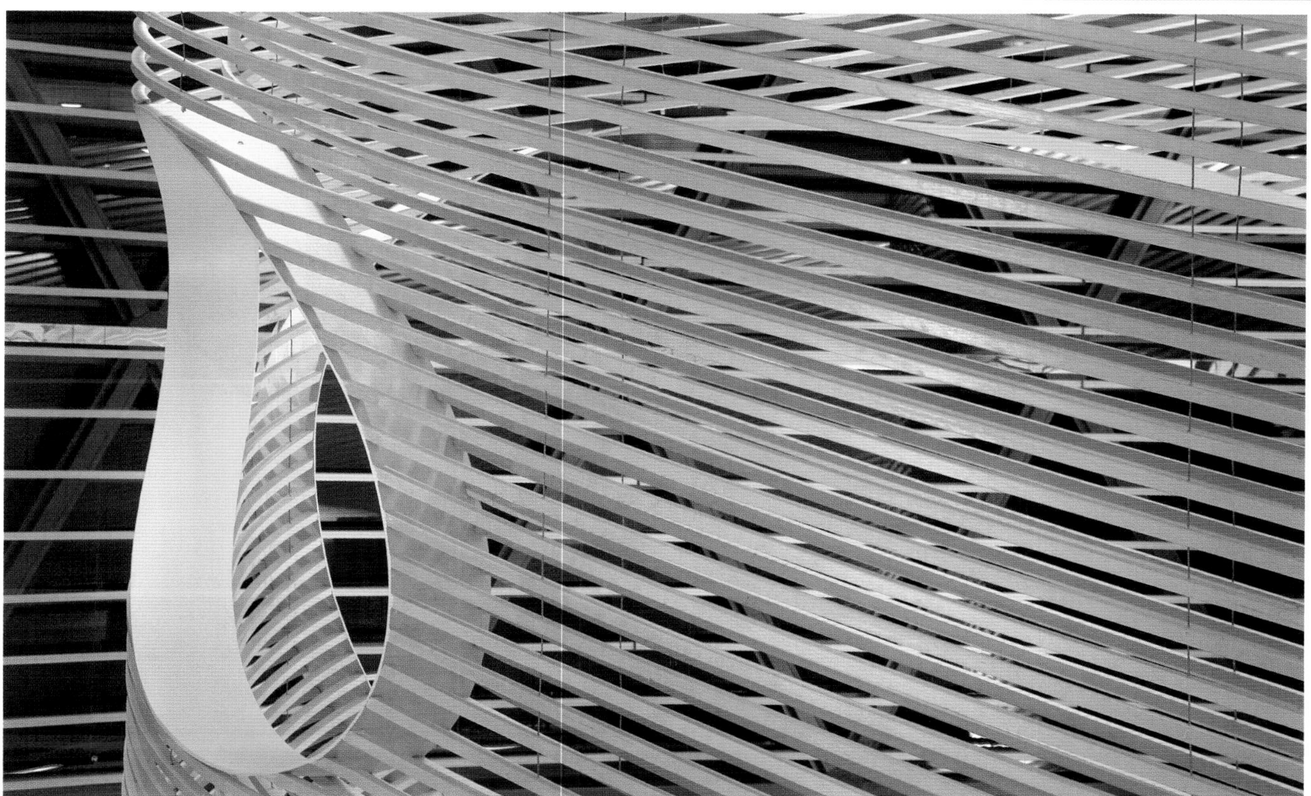

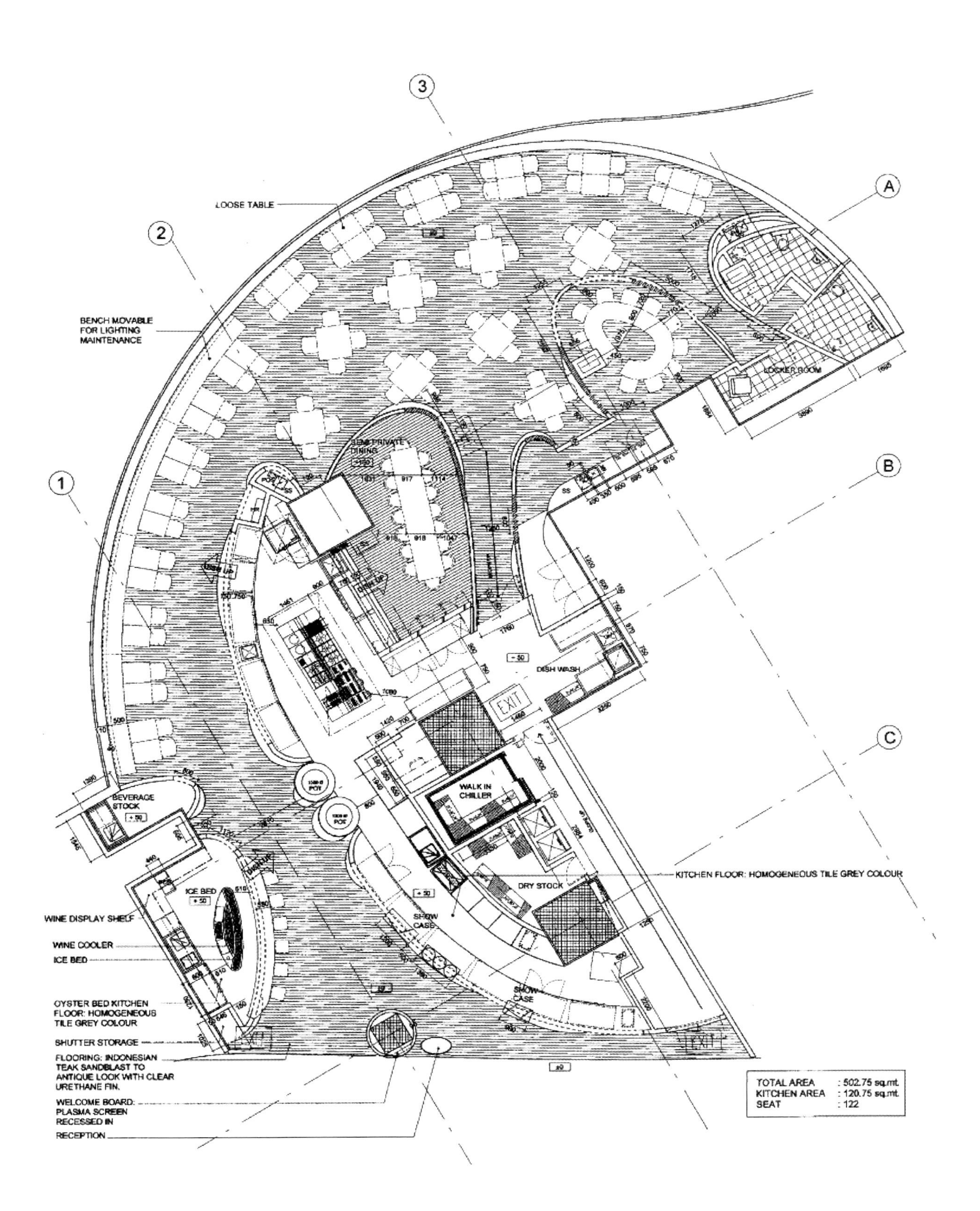
LOOSE TABLE
BENCH MOVABLE FOR LIGHTING MAINTENANCE
SEMI PRIVATE DINING
LOCKER ROOM
DISH WASH
EXIT
WALK IN CHILLER
BEVERAGE STOCK
ICE BED
DRY STOCK
SHOW CASE
KITCHEN FLOOR: HOMOGENEOUS TILE GREY COLOUR
WINE DISPLAY SHELF
WINE COOLER
ICE BED
OYSTER BED KITCHEN FLOOR: HOMOGENEOUS TILE GREY COLOUR
SHUTTER STORAGE
FLOORING: INDONESIAN TEAK SANDBLAST TO ANTIQUE LOOK WITH CLEAR URETHANE FIN.
WELCOME BOARD: PLASMA SCREEN RECESSED IN
RECEPTION
TOTAL AREA : 502.75 sq.mt.
KITCHEN AREA : 120.75 sq.mt.
SEAT : 122

HAMA

Designer: K-Studio
Design Company: K-Studio
Location: Glyfada, Athens, Greece
Area: 200 m^2
Materials: Epoxy Flooring, Black Steel Panels, Bamboo Canes and Flooring, Brass
Photographer: Yiorgos Kordakis

HAMA is a Japanese-Brazilian restaurant, located in the southern suburbs of Athens. The concept was influenced by the philosophy of the traditional Japanese house layout, reaching a balance between open space (oku) and closed internal space (ma). The design of the layout of Hama focuses open, social space near the entrance and bar, whilst natural bamboo screens enclose several small areas giving them privacy.
To achieve a gradient between open, social and intimate, private space the designers raised several individual dining rooms up above the ground floor, where, like glowing bamboo lanterns, they overlook the main space below, connected by suspended steel walkways. As if sitting high within a bamboo forest, diners are subtly screened from view yet remain connected to the ambience. These "lanterns" are stacked on top of each other, providing partially enclosed corners of space whilst maintaining the connection to the open dining area.
The lighting is directed and focused, bringing out the beauty of the bamboo and accentuating the contrast between its warmth and the dark steel and floor.

HAMA餐厅兼具日式和巴西风格，坐落在雅典南面的郊外。它的设计理念受到日本传统房屋布局的影响，即开放空间与封闭空间彼此平衡。HAMA餐厅的设计重点是靠近入口和吧台的开放式社交空间，同时，天然竹质屏风也间隔出几个包间，使客人享有私密性。为了实现开放空间与私密空间的自由转换，设计师将几个包间"提起来"，使它们看上去就像几个随风飘动的竹子灯笼一样，高出地面，通过悬浮的不锈钢连接过道，俯瞰整个空间。像是坐在一个空中竹林里，客人几乎不受他人的影响，悠然地享受着美食。这串"灯笼"为客人提供了相对独立的就餐一隅，同时又保持与室外餐厅的联系。

通过灯光的控制和聚焦，凸显出竹子的美丽，同时突出温馨的"灯笼"与暗色的地板以及不锈钢栏杆之间的反差。

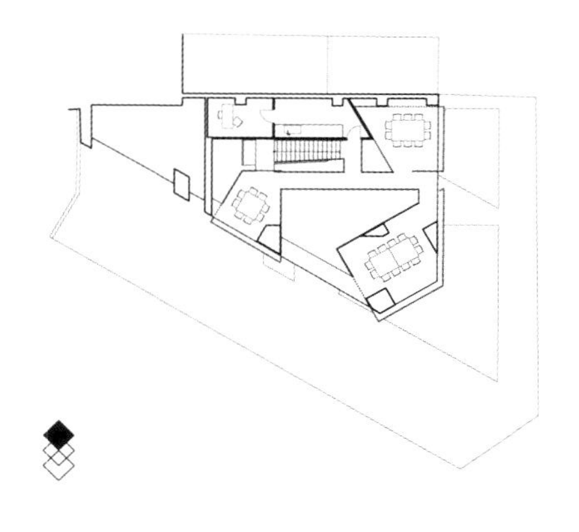

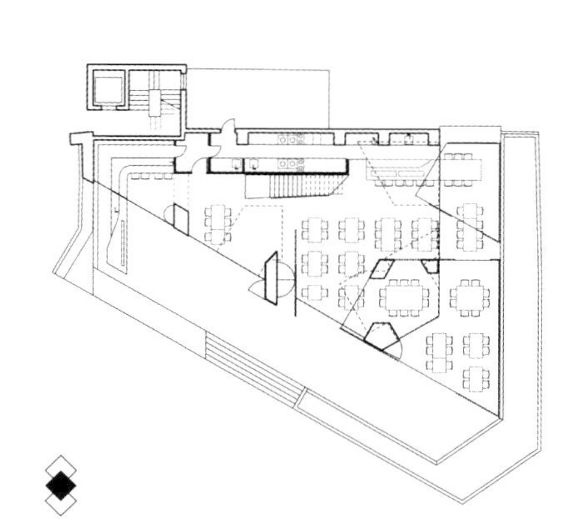

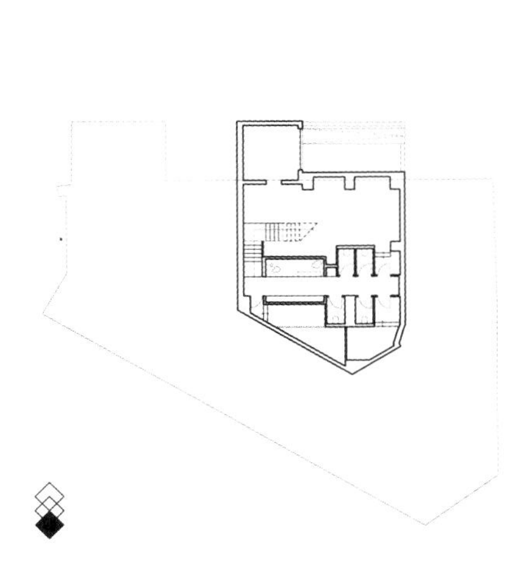

STOKEHOUSE

Designer: Pascale Gomes-Mcnabb
Design Company: Pascale Gomes-Mcnabb
Location: Melbourne, Australia
Photographers: Mark Roper, Sharyn Cairns

Design abstract

The brief from the client was to update and add panache to the Stokehouse, an iconic but faded dining institution on the foreshore at St Kilda beach Melbourne, without changing the core positive ambience.

To offer the guest a dining experience that was a departure from the old Stokehouse with its layers of disparate and worn elements accrued over bygone eras. And retain its beachside shack feel, a refreshed version that would appeal to a younger generation and yet avoid alienating their existing clientele.

Design challenge

The project context contained a demanding timeframe complicated by planning clearance on a heritage building, an intensive brief of requirements and exacting repairs and replacement of building elements on a tired old Edwardian teahouse.

The dining space had to segue from day into night when the ocean, beach and sky disappear and the focus becomes internal. The dining room becomes the star and imparts a sense of drama and occasion.

Creating two new distinct areas-a cocktail bar and private room that can be segregated and on other occasions flow into the main dining space, enlarging the kitchen and toilet facilities.

Design result

The design concept and newly-built environment brings a fresh beachside insouciant glamour to the upstairs dining room, new cocktail bar and terrace that shifts effortlessly from day into night.

The color palette is subtle verging on minimal. It refers to sparkling sea, golden sand, porphyry and pink brush strokes resonating the fading glow of sunset, salt, sun and bleached driftwood, beach bathing box stripes. At night the lighting is enveloping and sensuous, understated and harmonious.

Lux materials were chosen to give the space a discrete pared back opulence contrasting gold elements: brass bar, glowing columns and gold mirrors reflect surrounding vignettes and add the sunkissed glamour essential to an Australian beachside dining establishment.

设计摘要

业主的基本要求是翻新Stokehouse并添加一些华丽的因素，同时隐去“餐厅”元素，使之成为墨尔本圣基尔达海滩上的地标性建筑。

设计目的是为客人提供有别于老Stokehouse店完全不同的美食体验，在这里不见任何古旧的元素。保留其海滩休憩小屋的感觉，同时通过一个全新的视觉享受来吸引年轻人，但要避免伤害年老客人的感情。

设计挑战

时间紧迫，把一个传统的建筑彻头彻尾地翻修实属不易。

就餐区全天候开放，当夜幕降临时，内部的装饰尤显突出，在黑夜的衬托下熠熠生辉，如同一颗闪耀的星，亦幻亦真。

在主餐区设计两个截然不同的分区——鸡尾酒吧区和包房区，平时还可以合二为一，这样就可以扩展厨房的空间和增设卫生设施。

设计效果

设计理念和焕然一新的环境赋予了楼上餐厅一种全新的滨海风貌，优雅不俗；崭新的鸡尾酒吧和露台让人倦意顿消。

餐厅设计参考了波光粼粼的海面、金色的沙滩、斑岩和粉色的笔迹，与落日的余晖、盐滩、阳光和漂白的浮木、滨海浴箱上的条纹交相呼应。夜晚，灯光柔和而温馨，基调素雅且和谐。

选用的勒克斯材料，给空间增色不少，与金色元素，例如，铜色酒吧、明亮的立柱和金色镜子映射的装饰画形成鲜明的对比，呈现出一派少有的奢华，凸显了澳大利亚滨海餐厅的特色。

JAMIE'S WESTFIELD

Designers: Tim Mutton, Jo Sampson, Jordan Littler
Design Company: Blacksheep
Location: London, UK
Area: 610 m^2

Blacksheep has designed the latest "Jamie's Italian"restaurant for Jamie Oliver at the Westfield London shopping centre at White City, which the chef has called 'probably our most stunning site to date'. The "Jamie's Italian" restaurants are Jamie Oliver's first completely independent restaurant venture and aim to bring "what's best about casual dining" to the high street.

The restaurant exterior has a great L-shaped space (appropriate 100 m² in total) on Westfield's Southern Terrace, which offers al fresco dining beneath a canopy, suitable for year-round usage, with all restaurant doors and full-height windows able to be opened out onto the space in warm weather. The new, branded ice-cream van sits at the corner of the L-shape. The external area is bordered and demarcated by planters made from reclaimed scaffold boards, which are full of herbs, creating both a great smell and a sense of anticipation for the food to come.

The 610 m² interior is very long (55 meters from the main door to the door by the antipasti area towards the rear of the restaurant), with a particularly narrow central area, which wasn't ideal for seating. This gave rise to Blacksheep's ideas for the optimum space-plan and for zoned areas, from the main restaurant space ("Piazza") to a central "Market Place" area, without seating, which made a virtue of the narrow central space and where customers can see views of the kitchens and fresh pasta making or can linger over a bigger retail area than ever before, en route to either the toilets or else to the restaurant's rear dining space, The Back Room, a special, more intimate area with lower ceilings and a slightly more "bling" treatment.

Overall, the interior has a pared-down industrial feel, with exposed gantries, steelwork beams and columns, dark grey paintwork and lots of timber. It is spacious (with 4.2-meter ceilings in the first two zones), warm and accessible but doesn't pretend to be what it isn't. "This isn't in any way a pastiche of an old building", explained Mark Leib,"It expresses its modernity quite clearly."

T-SHIRTS £15=
S/M/L/XL
TEATOWELS £10=
A SET OF TWO DIFFERENT DESIGNS IN 100% COTTON.
JUCO BAG £7=
BIG & STRONG ENOUGH FOR ALL YOUR SHOPPING. 50% JUTE, 50% COTTON
COOK'S APRON £15=
ONE SIZE FITS ALL.
COFFEE MUG £8=
GREAT LITTLE MUG WITH MEASUREMENTS FOR THE PERFECT CUPPA.
OVEN GLOVE £12=
NAPKINS £12= (COMING SOON)
SET OF FOUR OF OUR SPECIAL "JAMIE'S ITALIAN" NAPKINS.

Blacksheep公司为Jamie Oliver设计了最新的Jamie Italian餐厅，位于White City区的伦敦Westfield购物中心。厨师将该餐厅称为“最令人震撼的约会场所”，它是Jamie Oliver第一个完全独立式餐厅，旨在将最好的休闲就餐场所带到这个繁华的商业大街。

该餐厅外面有一个大约100平方米的L形广场，正位于韦斯菲尔德购物中心的南侧露台，客人可以在天棚下面享受户外用餐，全年开放。天气暖和时，餐厅的所有门和落地窗户都可以打开，和这个广场连成一体。新的、名牌冰激凌车停在广场的一角。广场上摆放着回收的脚手板做成的花盆，里面种着很多草本植物，散发出浓浓的香味，营造出对食物渴求的氛围。

该餐厅面积为610平方米，空间呈长条形（从主门到餐厅后部开胃菜区的门有55米长），中央区域非常狭窄，并不是就餐的理想座位。这一点激发了Blacksheep公司的设计灵感，由此设计出了最佳的空间平面图和分区平面图，从餐厅的主空间Piazza区至中央的市场区（Market Place）不设置就餐座位，这样客人就可以通过狭窄的中央区域看到厨房的内部景象，也可以看到新鲜的意大利通心粉的制作过程，或许会在较大的零售区逗留一阵子，也有可能在去洗手间或去餐厅后面的就餐区途中稍作停留。餐厅后面有个更加私密、特殊的区域，里面有低低的天花板和不是很炫的装饰。

整体来看，内部装饰充满工业感，简洁、舒适，例如，裸露在外的门架、钢结构的横梁和柱子、深灰色的漆面和大量的木质装饰材料。内部非常宽敞（前两个区的天花板有4.2米高）、温暖，装饰风格亲切舒适、不做作。Mark Leib解释说：“这并非一个旧建筑的模仿品，而是现代感十足的新式餐厅。”

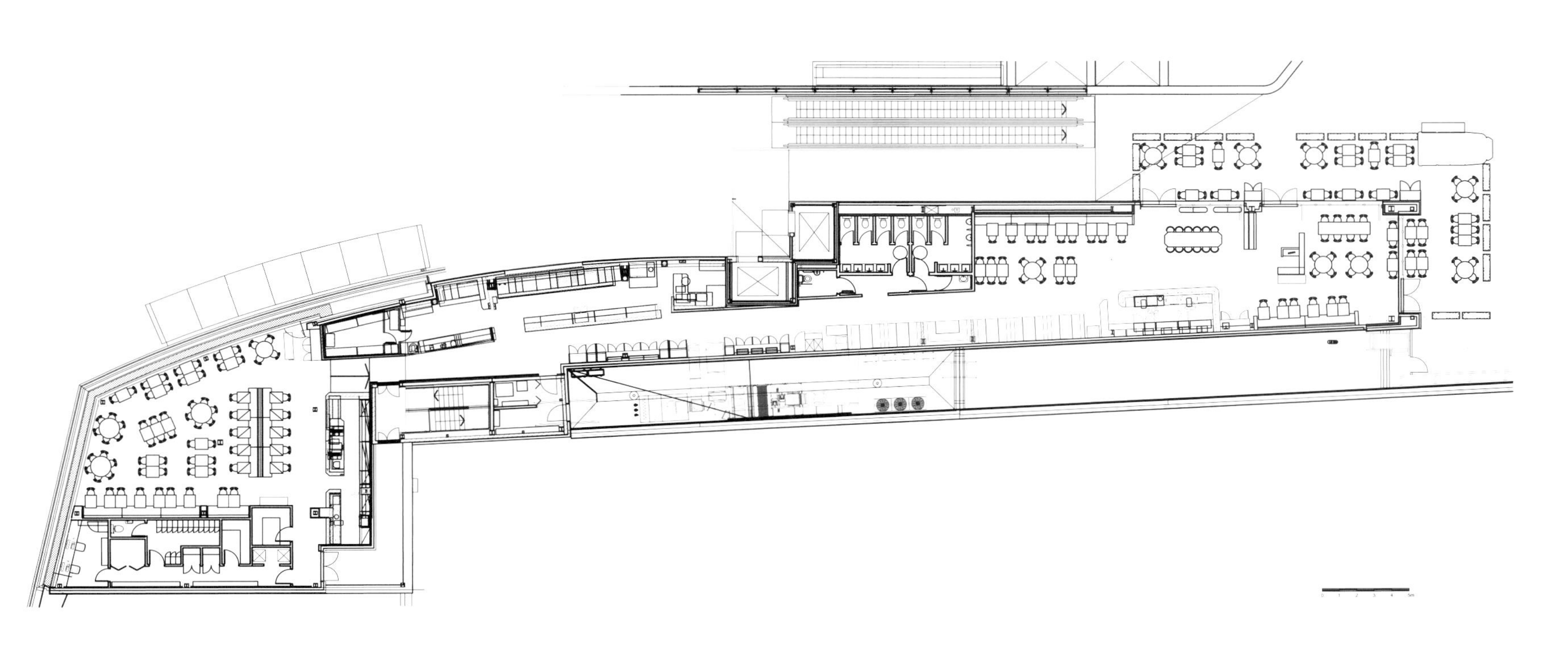

ALEMAGOU

Designer: K-Studio
Design Company: K-Studio
Location: Ftelia Beach, Mykonos Island, Greece
Photographer: Yiorgos Kordakis

The concept is holistic, with every element of the project telling a common story and coming together to create an inspirational yet laid-back atmosphere, perfectly suited to the site.
The menu revisits traditional, well-loved Greek recipes to produce simple but intriguing new dishes. The cocktails and the music selection also offer new combinations of favourite ingredients. The architectural design is an extension of the same approach, creating an exciting reinterpretation of tried and tested, traditional building techniques.
Taking inspiration from typical Cycladic architectural elements such as whitewashed, smooth-edged houses, dry-stone walls that blend into the scrubby landscape,practical, hardwearing screed floors,and natural reed-thatched roof insulation, the familiar textures are applied to contemporary, organic forms to create a unique character.
Added to the palate are the dominating natural conditions of the site: the strong winds that make this beach a surfer's paradise,as well as the burning 40-degree midday sun and the harsh, dry, rocks of the encroaching landscape. Rather than attempting to block the effects of these natural forces, the design welcomes them.
Natural reed thatch is used to create a 60cm deep, inverted field for a canopy that sways soothingly in the dissipated wind, allowing air to circulate and the space to stay cool. Throughout the day, dappled sunlight filters through the reeds, lighting and shading the space simultaneously. At night, down lighting continues to animate the canopy from within, creating a warm, intimate atmosphere for evening dining.
Beneath the canopy an un-interrupted topography of cool screed terraces flows gradually down from the restaurant to the sand, via the bar and lounging areas. Circulation is organized to create specifically designated areas for dining, drinking or relaxing yet their softly blurred, low-level boundaries let light and air flow naturally through the open space and allow continuous views across the beach to the sea and sunset.
The combination of these purposely designed and naturally occurring elements creates a multi-sensory architecture that sits in harmony with the environment, provides a natural, comfortable refuge from the elements ,and creates an exciting, sociable atmosphere.

设计理念注重全局，项目的各个元素相对独立又浑然天成，共同营造出活泼、轻松的氛围，完美地诠释着餐厅的风格。

菜单上呈现了人们十分喜爱的、传统的希腊菜谱，菜品简朴而又神秘。鸡尾酒和精选的音乐也是客人的最爱。统一的建筑设计理念贯穿始终，并且大胆地打破常规。

设计灵感源自典型的基克拉迪建筑风格，圆润的、经过粉刷的房子与石头墙融入到自然景观之中。实用的自流平地面与天然芦苇隔热顶棚，以现代工艺运用熟悉的传统纹理创造出独特的个性空间。

给这种风格锦上添花之笔是其独有的天然条件：强大的风势使得这里成为冲浪爱好者的天堂，还有正午40度的高温和干燥、粗糙的岩石，与设计意图简直就是天作之合。

以天然芦苇在空中搭建一个60厘米厚、可随风摆动的顶棚，这使得空气流通从而保证空间凉爽。白天，日光透过芦苇过滤制造出斑驳的光影，自然采光与阴凉兼得。夜晚，灯光穿过顶棚照射，被晚风吹拂摇动的灯火创造出温馨、亲切的夜间用餐环境。

顶棚下的自流平地面逐渐经过阶梯而过渡到沙滩，形成酒吧和多功能休闲区域，Alemagou有特定的用餐区，而饮酒和社交则在一个界限较为模糊的区域，光线和空气在这个开放的空间中更加自然地流通，客人穿过海滩就可以欣赏到大海和日落。

这些独特的设计融合了自然元素，创造出一个多感官的、完美融入环境的建筑，一个自然、安逸的世外桃源和社交场所。

PARADIS DU FRUIT

Designer: Philippe Starck

This is a restaurant featuring fruit products, which locates in George V Avenue ,Paris, with an area of 350 m². It is separated into the main dining area, balcony, terrace and other functional areas. The restaurant can accommodate a total of 200 people.

Early in the design, the designers and restaurant owners make an agreement with each other, that is, using the wood as the main material to create a tropical rain forest-style environment. By doing this, the designers try to highlight the restaurant respected healthy, organic and natural cliet conception. Surrounded by the parquet floors and mahogany wainscoting, the entire space has created a warm and soft feel. In contrast, it appears that the design of the bar area is much shaper. The bottom of the bar adopts white floor, which gives a natural division of the territory exclusively bartender. The combination of stainless steel bar and mirror-coated back of bar brightens the entire area as well as making the fresh fruits and vegetables in the counter seem more mouthwatering.

A few stainless steel columns are also eye-catching in the dining area. With close watch, guests will find that designers make it into the shape of the trunk. The roof of the chandelier is from the artist Aristide Najean. Dendrite general shape and stainless steel trunk get together to create a "common sense forest" style.

In addition, the designers also reflect fruit elements in the details, such as, using glass texture to express the texture of flesh. On the walls around the restaurant, hanging with a total of 14 surface plasma display with 2.5 meters x 1.5 meters,which loops video clips by the designer as a creative art. And when the restaurant holds parties or other activities, these screens become important props.

MELON SUCRIN, QUINATA, MANGUE
CALAMONDIN, CACAO, NOIX, CITRON
OLIVE D'AUTOMNE, FRUIT DE LA PASSION
PAMPLEMOUSSE, POMME CANNELLE
FRUIT DE LA PASSIFLORE, FRAISE
FRUITS DE FUCHSIAS
MELON, PAPAYE, FIGUE, FRUIT DE SEBESTIER ORANGE, KIWI
FRUIT DE L'ARBRE A PAIN, GRENADILLE, POMELO
PIMENT CARRÉ, KIWANO, LITCHI, MAYPOP, TOMATE
GROSEILLE, MELON CANTALOUP
MANGOUSTAN, CORME, COING, KIWAI
PHYSALIS, CURUBA ROUGE, GRENADELLE
NOIX DE COCO, POMME LIANE, NOIX DE MACADAM
CEDRAT, CERIMAN, PASSIFLORE

ERISTOFF

这是一家以水果制品为特色的餐厅，位于巴黎的George V大街，占地面积达350平方米。分为主就餐区、包厢、露台等功能区域，总共能够容纳200人。

在设计之初，设计师与餐厅老板达成了统一意见，即以木材为主要材料来营造热带雨林风格的环境，从而突出该餐厅推崇的健康、有机、天然的饮食概念。在镶木地板及桃花心木护墙板的包围中，整体空间产生了一种温暖而柔和的感觉。相比之下，吧台区域的设计则显得犀利许多，吧台底部采用白色木地板，自然划分出专属于bartender的领地，镜面包裹的不锈钢吧台及结合了灯光背板的酒柜将整个区域提亮，令柜台里的新鲜蔬果看上去更加娇艳欲滴。

同样抢眼的还有就餐区的几根不锈钢柱子，仔细观察就会发现，设计师将其塑造成了树干的形状，而屋顶的吊灯则出自艺术家Aristide Najean之手，枝蔓一般的卷曲造型与不锈钢树干共同营造出“铿锵森林”的风格。

除此之外，设计师还将水果元素体现在细节之处，例如，用玻璃的纹理来表达果肉的质感。而在餐厅四周的墙上，一共挂着14面2.5米x1.5米的等离子显示屏，循环播放由设计师作为艺术指导创作的影像片段，而当餐厅用来举行派对或者其他活动时，这些显示屏就成为重要的道具。

AVVOCATO PENTITO

Designers: Jaime Gaztelu, Mauricio Galeano, Iñigo Esparza, María Asún Osés
Design Company: James & Mau Architects
Location: Pamplona, Spain
Area: 420 m^2
Photographer: Pablo Ausucua

The space is divided in three sections (lounge, coffee-shop, restaurant), on the three levels. All the three levels follow a unity of color and materials, (Monochrome pictures, cement, glass) creating a modern and fashionable feel. Despite this coherent unity each section manages to enhance its own specific ambiance: the lounge is set in the basement and a strong sense of cosines is created with the black leather sofas, apparent original rock walls and pop drawings; on the street level is the coffee shop, where the space has been sectioned like slices of salami thanks to integrated Plexiglas modules that work as tables and side illuminated walls from one end to the other of the space. These modules create the impression of a virtual passage going from the entrance in direction to the bar, obliging guests walking through to be seen and to see. One the first floor is the restaurant area, with a more traditional and elegant feel to it.

- Structure: Reuse of existing structure based on 60-cm-thick stone load bearing walls and traditional timber slabs reinforced punctually on metal structure.
- Basement Ground Slab: Reinforced concrete over draining water pumped system.
-Partitions: Gyssum board and Dry Wall system.
- False Ceilings: Gyssum board and Dry Wall system according to acoustic regulations.
- Floor Finishes: Epoxi painting on light grey color. Ceramic floors in restrooms and kitchen.
- Wall Finishes: Plastic painting on light grey color as main base, Serigraphed Plexiglass table modules on the ground floor, Ceramic walls in restrooms and kitchen.
- Facade: Sand washed existing stone and 2-mm-thick metallic sheet oven painted on light grey.

该空间被分割成休息大厅、咖啡店和餐厅，每个部分各占一层。三层色调一致，选材相同（黑白图片、水泥、玻璃），营造出现代、时尚之感。在追求整体和谐一致的前提下，各个部分又要突出特点：休息大厅设在地下室，呈弧形摆放的黑色皮沙发、粗糙的石头墙和喷墨画营造出一种强烈的流线感。地上一层是咖啡店，大量的橡胶板块用在桌子上和通体影壁灯墙上，这使得这一层被分割成众多小间，看上去就像一串串腊肠。这些模块营造出一个虚拟长廊，从入口处一直通向酒吧，吸引着客人穿廊而过，欣赏别人也展示自己。二楼是餐厅，装饰传统而优雅。

结构：在原本60厘米厚的石头墙和传统木质板材的基础上，重点搭建了金属框架。

地下室的地面砖：混凝土砖铺在排水系统之上。

隔墙：采用Gyssum板材和Dry Wall系统。

装饰天花板：根据声学原理，采用Gyssum板材和 Dry Wall 系统。

地板踢脚线：采用Epoxi浅灰色喷漆。卫生间和厨房采用Ceramic地板。

墙裙：将浅灰色防水漆作为基本色调。一楼摆放着Serigraphed Plexiglass桌子。在卫生间和厨房墙面上铺设了陶瓷砖。

正面：水磨沙石和2毫米厚的金属铁皮均采用浅灰色喷漆。

CACIQUE

VESU RESTAURANT

Designer: Arcsine Architecture
Design Company: Arcsine Architecture
Interior Finishes and Furniture: Bellusci Design
General Contractor: Terra Nova Industries
Structural Engineer: Gregory Paul Wallace, SE
Mechanical, Electrical and Plumbing Engineering: Encon Consulting Engineers
Kitchen Design: JM Design & Associates
Lighting Design: Caprice Carter Lighting Design
Location: Walnut Creek, California, USA
Photographer: Sharon Risedorph

Arcsine Architecture and Bellusci Design transformed a piano store with little curb appeal into a vibrant and engaging restaurant, fitting the bustling downtown Walnut Creek. The clients sought an inviting, modern environment to complement the globally inspired, locally sourced menu by Executive Chef Robert Sapirman. Situated on a competitive street corner, the design needed to grab attention and engage passers-by.

Arcsine wrapped the building with a contemporary, wood-panel façade, complimented by a curving wall of frameless glass that entices pedestrians to look in and feel as if they are already inside. An Arcsine-designed canopy of FSC Certified Eucalyptus leads patrons from the entrance to the lounge and extends all the way to the back bar. Bellusci Design continued the contemporary motif, selecting furniture and interior finishes in a warm, monochromatic spectrum with plenty of texture. Combining these elements, Vesu wields a commanding presence—one that draws guests, welcoming them into an open, communal space that is simultaneously sophisticated, comfortable and warm.

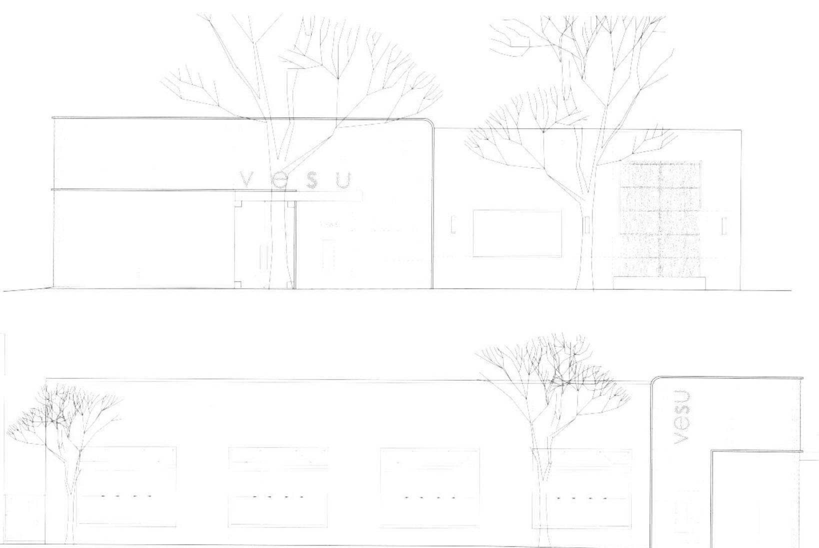

vesu
vesu

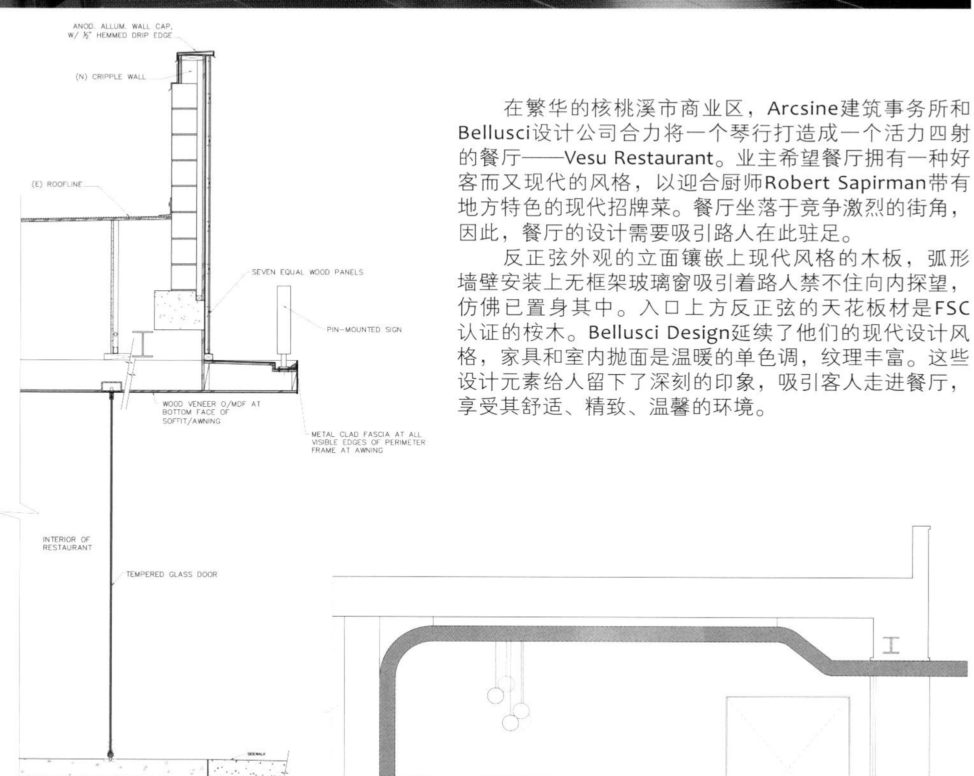

在繁华的核桃溪市商业区，Arcsine建筑事务所和Bellusci设计公司合力将一个琴行打造成一个活力四射的餐厅——Vesu Restaurant。业主希望餐厅拥有一种好客而又现代的风格，以迎合厨师Robert Sapirman带有地方特色的现代招牌菜。餐厅坐落于竞争激烈的街角，因此，餐厅的设计需要吸引路人在此驻足。

反正弦外观的立面镶嵌上现代风格的木板，弧形墙壁安装上无框架玻璃窗吸引着路人禁不住向内探望，仿佛已置身其中。入口上方反正弦的天花板材是FSC认证的桉木。Bellusci Design延续了他们的现代设计风格，家具和室内抛面是温暖的单色调，纹理丰富。这些设计元素给人留下了深刻的印象，吸引客人走进餐厅，享受其舒适、精致、温馨的环境。

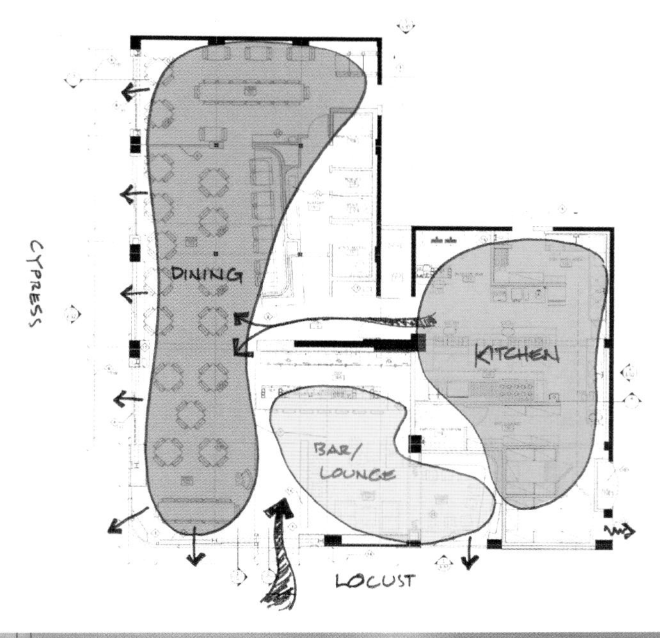
DINING
KITCHEN
BAR/
LOUNGE
CYPRESS
LOCUST

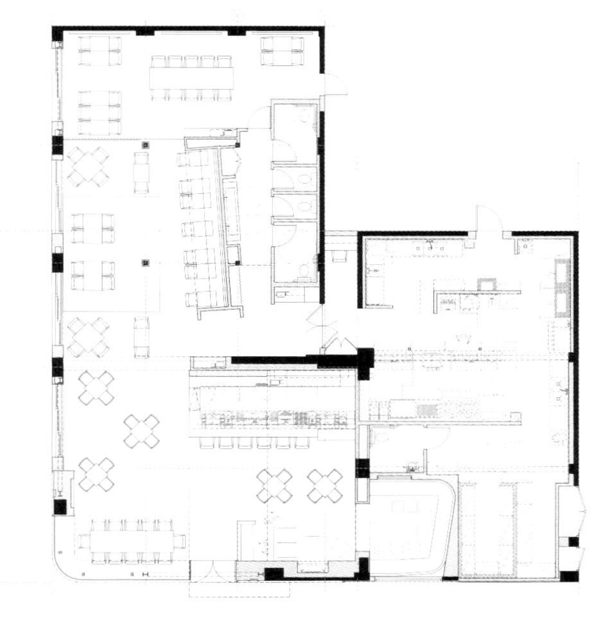

vesu

M
W
W
vesu

DESIGNERS

David Rockwell

Long before turning his attention to architecture, David Rockwell harbored a fascination with immersive environments. Growing up in Chicago, Illinois and Deal, New Jersey, and Guadalajara, Mexico, Rockwell was a child of the theater – his mother worked as a vaudeville dancer and choreographer and cast him in community repertory productions. Rockwell brought his passion for theater and an eye for the color and spectacle of Mexico to his architecture training at Syracuse University where he received a Bachelor of Architecture degree and to his studies at the Architectural Association in London. These formative influences continue to have a strong impact on his practice.
Rockwell Group is an award-winning, cross-disciplinary architecture and design practice. Based in New York City with a satellite office in Madrid, Rockwell and his 140-person firm focuses on a diverse array of projects that range from hotels to hospitals, restaurants to airport terminals and Broadway set designs to consumer products.

SHH

SHH is an architects' practice and interior and branding design consultancy, formed in 1992 by the three principals of the practice: Chairman David Spence, Managing Director Graham Harris and Creative Director Neil Hogan. With a highly international workforce and portfolio, the company initially made its name in ultra-high-end residential schemes, before extending its expertise to include leisure, workspace and retail schemes. The company's leisure CV includes McDonald's breakthrough Oxford Street flagship restaurant, winner of 6 international design awards; new concept Teaspoon tea rooms in Russia; school dining areas for The School Food Trust and award-winning leisure centres for LA Fitness and 37°. SHH's work has appeared in leading design and lifestyle publications all over the world, including VOGUE and ELLE DECO in the UK, Artravel and AMC in France, Frame in Holland, Monitor in Russia, DHD in Switzerland, ELLE DECO in India, Habitat in South Africa, Contemporary Home Design in Australia, IDS in Malaysia and Architectural Digest in both France and Russia, with over 100 projects also published in 50 leading book titles worldwide.

Koichi Takada

Koichi Takada studied and graduated from the City University of New York and the Architectural Association in London. As an Associate Director at PTW Architects in Sydney, Koichi worked on a diverse range of projects in Australia, Japan, China, Middle East in Dubai, Abu Dhabi. In 2009, Koichi established Koichi Takada Architects in Sydney, and is currently working on various projects in Australia, Japan, and Europe. Koichi has won key personal awards including the Grand Prize from Central Glass International Architectural Design Competition, SD Review Award, Australian Premier Award, Australian Interior Design Awards, and also has taught at the University of Technology Sydney (UTS), Royal Melbourne Institute of Technology (RMIT) in Melbourne.

SUE Architekten ZT KG

Three partners: Architekt DI Christian Ambos, Architekt DI Michael Anhammer, Architekt DI Harald Höller
www.sue-architekten.at
office@sue-architekten.at

Jim Hamilton

Jim is Creative Director at Graven Images and leads the Hotels & Leisure team. During his 19 years with the studio, he has helped to create inspiring places that are both popular and practical.
Jim is currently working on a number of hotel projects in the UK, Europe and the US. He is leading the design team for the Arne Jacobsen Radisson Blu Royal Hotel in Copenhagen and a range of projects in the US for Carlson Hotels.
Closer to home, recent projects he has worked on include the design of Tigerlily, a multi award-winning boutique hotel for Montpeliers in Edinburgh, and Blythswood Square, Glasgow's newest five-star boutique hotel and spa.
He frequently lectures on design issues at Universities and Colleges, including Glasgow School of Art, Glasgow Caledonian University and Duncan of Jordanstone College of Art and Design, and for the Interior Design Institute. He is also a member of the Glasgow Metropolitan University Design Review Committee. Jim is an external examiner for the BDes in Interior Design at Duncan of Jordanstone College of Art and Design.

Bluarch Architecture & Interiors

Bluarch Architecture & Interiors in New York, NY is a private company categorized under Architects. Current estimates show this company has an annual revenue of $500,000 to $1 million and employs a staff of approximately 1 to 4. Companies like Bluarch Architecture & Intrrs usually2 offer: David M Schwarz Architectural Services, Architectural Drawing Services, Architectural Engineering Services, David M Schwarz Architectural Services Inc and Architectural Development Services.

Ryntovt Design

The beauty of exploded bud, respect from life to life, the soul of the place, the wisdom of the forest, love of the earth – these are the basis of our work, our relations with nature and world. The fundamental concept is to walk on the Earth whispering.
Everyone is the part of one live entity, and should respect and love it appropriately, this is entire system, spirit and materiality are bounded together, part reflects the whole and there the human being is a part of the world with each molecule. Everyone creativity is based on this philosophy, sense of the work and attitude to it. 15 –year experience allows to affirm that – this ideology is future.
Wish to spend the present day in the feeling of full consciousness allows to enlarge the edges of apperception and action.
Each object is full of spirit, based on the culture of manufacturing, respect to nature, gratitude to the material, which gives us the possibility to implement senses, which are so important for us.
The object which is the result of our creativity, first of all is a storage of ecoculture and spirituality.

Urbantainer

Urbantainer is an architecture and design firm that creates unique atmospheres and spaces for subculture inhabitation within the city. Various social actions and behaviors are continuously taking place within the built urban environment. They seek this social activity and offer it a presence. Their designs are immediately inhabited by events for people to meet, interact, and enjoy. We give subculture a physical presence within the city.

Creneau International

Creneau International is a global agency that connects brands, interiors and consumers through design concepts.
Since 1989, they have established a strong reputation as visionary and trendsetting designers. Their ideas of the anti-ordinary make brands flourish, products grow and businesses boom. Their added value is to provide full service: from all-round consultancy to concept development, from logo and packaging design to interior design, from implementation to franchise follow-up.
Whether a challenger brand around the block or an iconic brand spanning the globe: they help you reach the audience. In shops, shops-in-shops, hotels, restaurants, bars or nightclubs as well as offices, showrooms, exhibition stands and events.
FOR all , they tell the story on the moon. At Creneau International, they know the success is written in the stars. Like they say: hac itur ad astra.

Tao Thong Villa + Process5 Design

Tao thong villa was established by Hideki Kureha in April 1996.
He does wide range of work, wedding photography, commercial photography, graphic design, interior design as well as architectural work.
His main office is located in Aoyama, Tokyo, brunches are in Nagano, Matsumoto, Osaka, and Himeji.
Process5 Design was started as Process5 by Noriaki Takeda and Ikuma Yoshizawa in 1999.
The office was established as Process5 Design in 2009 in Osaka, Japan. Process5 design design architecture, interior, and graphic which valued the concept regardless of use.

Kuadra

Kuadra was established in 2001 by Andrea Grottaroli and Roberto Operti. We are a young and multitalented firm providing architectural design (civic, commercial, residential), industrial design and furnishings, interior design (residential, commercial and exhibition stands). They also offer graphic design solutions (from an integrated image for organizations and businesses, to the development of company logos, brands and internet sites). In addition to our professional services for clients, they also participate in many architecture and design competitions, winning many commendations and prizes for the entries.

James & Mau Architects

James & Mau, an architecture and design studio created by Spanish Architect Jaime Gaztelu and Colombian Architect Mauricio Galeano. They have produced projects in Chile, Colombia and Spain going from architecture of buildings and uni-familiar houses to interior architecture and design. James & Mau are also creators and partners of Infiniski, (www.infiniski.com) company specialized in Green construction and architecture, using mainly recycled and non polluting materials.

Yuhkichi Kawai

1967 Born In Tokyo, Japan
1992 Graduated Keio University Law dept.
1997 Graduated Kuwasawa Design Institute Space Design Dept.
Enter Super Potato Co.,Ltd.
2003 Establish Design Spirits Co.,Ltd.
2005—2006 Gold Award of Malaysia Society of Interior Designers
Gold, Silver & Honorary Mention Award of InteriCAD Interior Design
2009 Silver Award of Japanese Society of Commercial Space Designers
The Great Indoors Award 2009, Frame Magazine
Nominated for the Best of Year Award 2009, Interior Design Magazine

Hernandez Silva Arquitectos - Evo

http://www.hernandezsilva.com.mx
Members:
Name: Arch. Jorge Luis Hernández Silva
University: Iteso
Firm: Hernandez Silva Arquitectos
Name: Arch. Diego Vergara
University: U de G
Firm: Evo Arquitectos
Name: Arch. Lorena Ochoa
University: Iteso
Firm: Evo Arquitectos

Yasumichi Morita

Glamorous Co., Ltd.
Yasumichi Morita was born in Osaka, Japan in 1967.
He debuted as an interior designer at the age of 18 when he designed a bar "COOL" in Sannomiya, Kobe.
After working freelance for several years, Morita was awarded the position of chief designer at Imagine Inc., the noted design consultants.
In the wake of the Kobe Earthquake of 1995, Morita was determined to become independent.
In the following year, he set up the Morita Yasumichi Design Office which was restructured and renamed Glamorous Co., Ltd. in 2000
Morita participated "london 100% design" in 2000 and 2001 and won the IIDA (the International Interior Design Association) Will Ching Design Award for a national project in 2001. He also won the Hospitality Design Award hosted by the American magazine "Hospitality Design" for his international debut project.
Since then, he has successfully broadened his appeal to New York, London, Hong Kong, Shanghai and other major cities.

LW Design Group

In 1999, Lars Waldenstrom joined forces with Morten Hansen and Jesper Godsk to found LW Design Group as an interior design consultancy.
The principal ambition for LWD was to establish itself as a multi-disciplinary design studio whereby clients could benefit from a uniquely holistic approach, having all their architectural, engineering and interior design needs met by one firm. In line with this vision, Colin Doyle joined LWD in from RMJM in 2003 to head up a new architectural division and Finn Thielgaard established the engineering arm of LWD in 2006. What began as the vision of three men now has a staff of over 80 people based in the Design House in Media City and an architectural office that has been operating in Auckland since 2007.
Renowned for its unified design, functional space planning and contemporary style, LWD is today one of the region's premier commercial firms. Its portfolio encompasses 45 hotels and 160 restaurants and bars to date, counting some of the worlds top brands among its client's lists, including Le Meridien, Rotana Group, Raffles, Hyatt, Emirates Airlines and Jumeirah International.

Blacksheep

Blacksheep was set up in 2002 by Tim Mutton and Jo Sampson because of their shared passion for design. The agency they envisioned would refuse to offer generic, "house style" design, but would fly the flag instead for design that didn't play by the rule books – and which had the vision and ambition to set them and their clients apart.
They believe in holistic solutions, because people experience spaces and brands holistically. To achieve a space that really performs, They build multi-disciplinary teams of architects, interior designers and branding and graphics specialists, who understand everything from the impact of a great name and a great logo to the best way a brand can be interpreted and experienced three-dimensionally. The Blacksheep team – as cosmopolitan and international as their portfolio - includes designers who excel at concepts and designers who excel in detailed design and implementation, because they believe that the most luxurious spaces are defined both by the confidence of broad brushstrokes and by the tiniest and most surprising detail.

Tjep.

Tjep. works in a diversity of fields: from a Tiara for Princess Maxima to a nest consisting of giant rubber branches as a sofa. From the development of champaign popcorn for Volkswagen to complete interiors and identities for restaurants, schools and shops. Tjep. was founded by Frank Tjepkema, such a unpronounceable name required a design intervention in itself, which resulted in Frank's agency name: Tjep.
Tjep. consists of a small team of highly motivated designers based in Amsterdam and working for companies such as Droog, British Airways, Camper, Heineken and Ikea.
In every project the objective is to challenge conventional views, every project starts with a 'why?'. The execution, on the other hand, should be a delicious cocktail combining innovative ideas with functionality and visual elegance. Ultimately we are designing to add a little quality, energy and amazement to the world.

Klm-Architects

Klm-Architects realize innovative interior design, remodeling, renovation and new construction from offices in Berlin and Leipzig.
Experienced in design and realisation the architects have a focus area on high-quality solutions for hotels, guest houses, apartments and premium houses. The scope of klm services span all tasks from partial renovation of guest rooms, the redesign of individual suites up to the complete renovation and refurbishment of an entire hotel complex.
Second focus is the design of dental surgeries, medical centers and hospitals. Together with a network of experts and trusted specialists klm-architects cover all skills and disciplines which are necessary to make the solution become real.
The design philosophy of klm-architects is the creation of a form appropriate to place, location and environment as well as to the users wellbeing, through proportion of space, uniqueness of shape and a positive perception.

Pascale Gomes-McNabb

Pascale Gomes-McNabb has excelled in creating interiors for the restaurant ventures she and partner Andrew McConnell have opened. The interiors work not from over design, but from having just the right sort of whimsy, the right degree of comfort and ambience to match the level of love and attention to detail provided by the food. Cumulus Inc is yet another example of the winning combination of thoughtful design and food.
Pascale is an architect with a hospitality bent. She studied Architecture at RMIT and completed a bachelor of Architecture at Melbourne University. While studying architecture she also worked in hospitality and has now over 20 years experience. Juggling both interests led her to work in Sydney, London, Chicago, Hong Kong and Shanghai, before settling back in Melbourne 9 years ago.
In 2000, she designed and set up the restaurant, diningroom 211 with Andrew McConnell, her husband. They have just completed and opened their latest venture Cutler and Co in Fitzroy. She also designed and co - owns Cumulus inc in Flinders Lane, established in 2008.

Aidlin Darling Design

Aidlin Darling Design bridges the demands of artistic endeavor, environmental responsibility, functional pragmatics, and financial considerations. As a multidisciplinary firm, we believe that innovations discovered through the process of design and construction can be applied to projects of any scale, use, or purpose. The studio has a broad focus including institutional, commercial and residential architecture as well as furniture and interior design.
Partners Joshua Aidlin and David Darling have cultivated a team that strives to deliver extraordinary, responsible and innovative design. Their approach is client and site specific, and questions conventional assumptions. A collaborative process with clients, consultants, fabricators and builders allows an open and impassioned exploration and enables a clear understanding of appropriate solutions. In each project, we seek to uncover an inherent spirit of place and interpret constraints as catalysts for performative design. The individual character of each project emerges through poetic spatial relationships, material richness, and exacting detail.

Michael Young Studio

Michael Young Studio was founded in London in 1994 and incorporated in Hong Kong January 2006 by Mr. Michael Young, the studio is considered to be the most formidable presently in Asia and is bold enough to state that it is responsible for designing icons for its clients that live a lifetime, have won awards and are presented in museums globally. The key objective and mission of the company is to provide exclusive and quality design services within furniture, product and interior design markets. Michael Young's studio has a very unique environment to work along side with and is one of a kind in Asia. The studio is specialist in creating the connection of modern design and technical abilities of the local industry, in this case Chinese. Rising from the momentum of collaboration with Chinese in–dustrialists, the studio captures the strengthening ties between local industry and design, and exemplifies the skills of Asia industry and manufacturing.

Digital-Space

Digital-Space – founded 12 years ago by its owner and principal designer Eyal Shoan - is a multi-disciplinary design studio - specified in interior design, architecture, product design and virtual design.
The studio involved in various projects worldwide: hotels, yachts, residence, furniture, products design and multimedia productions. One of the recent projects of the team is the interior design of "Westminster Bridge Park Plaza Hotel", 1100 rooms trendy hotel in London. Currently the studio team works on the interior design of art hotel in Amsterdam and plans residence project in Phuket, Thailand.

SAQ

SAQ is a conceptual and interdisciplinary design agency specialized in developing spatial sceneries and concepts.
The practice relies on a broad range of competencies where architects, interior designers, urbanists, video-artists and graphic designers team up according to the specific orientation or necessities of each project. SAQ believes strongly in co-operation and regularly invites professional experts or companies to participate in the materialization or the elaboration of an idea.
For SAQ scale is no parameter. Intensive research, sketching, simulation, and dialogue are the fundaments of an intensive creative process leading to proposals for both small scaled designs and macro-planning.
The subjects on which SAQ is asked to ponder diverge from the common architecture commissions: the studio can be asked to contribute concepts for marketing strategies as well as for visual animations for public events.

Philippe Starck

The thousands of projects - complete or forthcoming - his global fame and tireless protean inventiveness should never distract from Philippe Starck's fundamental vision: Creation, whatever form it takes, must improve the lives of as many people as possible. Starck vehemently believes this poetic and political duty, rebellious and benevolent, pragmatic and subversive, should be shared by everyone and he sums it up with the humour that has set him apart from the very beginning: "No one has to be a genius, but everyone has to participate."
His precocious awareness of ecological implications, his enthusiasm for imagining new lifestyles, his determination to change the world, his love of ideas, his concern with defending the intelligence of usefulness – and the usefulness of intelligence – has taken him from iconic creation to iconic creation.
Inventor, creator, architect, designer, artistic director, Philippe Starck is certainly all of the above, but more than anything else he is an honest man directly descended from the Renaissance artists.

K-Studio

Since 2002 K-Studio has built up a varied portfolio of projects, believing that encouraging creative experimentation and fresh thinking leads to exciting architectural experiences on every scale and in every aspect of life.

Concrete Architectural Associates

Concrete architectural associates is founded in 1997.
The present director of Concrete Architectural Associates, Rob Wagemans, was born in Eindhoven on 13th of February 1973,he is a Master of Architecture Utrecht.
Erik van Dillen (at this moment, he is just creatively involved with Concrete), interior architect, was born in de Bilt on 27th of April 1960, catering industry skills in the kitchen, painting restorer.
Concrete originally was founded by Rob Wagemans, Gilian Schrofer and Erik van Dillen. They met each other by a not realized project, a head office in Amsterdam for Cirque du Soleil. Gilian Schrofer left concrete in 2004 to start his own company.
Concrete Reinforced is founded in 2006 by Rob Wagemans and Erikjan Vermeulen (present co-director of Concrete Reinforced). Erikjan Vermeulen is a Master of Architecture. He worked for different architects to start his own company in 2003.

Dan Pearlman

Dan Pearlman is a strategic creative agency, working interdisciplinary across the areas of brand and leisure, strategy and implementation.
The agency differentiates itself through its holistic approach, combining strengths from twelve different fields: Branding, Research & Innovation, Internal Branding, Brand Experiences, Fairs & Events, Retail, Visual Communication, Motion, Public Relations, Hospitality, Leisure and Zoo.
40 people contribute to the development and implementation of holistic solutions, enabling 360° Communication. If shop design, architecture for fairs and exhibitions or 3D Visualization – Dan Pearlman provides everything from the initial concept to the final execution under one roof.

II BY IV Design Associates

Founded in 1990, II BY IV Design Associates is one of Toronto's top four interior design firms and recognized in the World's Top 50. A boutique firm, we are known for cost-effective creativity and for pushing design excellence goalposts to support our clients with the best quality interiors we can. As a welcome consequence, we have been able to cultivate enduring relationships with a roster of impressive clients, many of them global brands. To date, II BY IV has earned more than 220 awards for outstanding design, worldwide.
Known for our accomplishments, we have come to be respected by our clients and industry for our innovation and attention to every detail.

Nemaworkshop

Nemaworkshop is a team of architects, designers, and thinkers who create spaces which are conceptually innovative and highly sensitive to cultural and social contexts. The studio approaches projects through research and collaborative brainstorming wherein ideas are discussed and reworked until the team emerges with a cohesive concept. The process is a non-linear approach, adhering to the conviction that good ideas can come from unlikely places.
Anurag Nema is principal of nemaworkshop. With 20 years of experience in hospitality design, he has completed over forty projects. Holding architectural degrees from universities in India and the United States, Nema was project manager at Rockwell Group and Partner at Studio Gaia before founding nemaworkshop in 2004. He has won several awards including the Design Award for Best Restaurant from Travel + Leisure for Delicatessen.

Katsunori Suzuki

Katsunori Suzuki is an Interior designer, born in 1967, Tokyo, Japan.
He is the president of Fantastic Design Works Co., Ltd.
Though our works are wide-ranging, we mainly design restaurants and dining bars, including club lounges and boutiques.
They figure out unique qualities as a business, and then produce a dramatic space with different design techniques.
Their work gives guests a sense of excitement and a vivid impression. It's their joy to offer these surprising spaces to out customers.

FCJZ

FCJZ is a leading architecture and design office in China. The practice was established in 1993 by Yung Ho Chang and Lijia Lu. FCJZ is internationally-honored with most prestigious recognitions in the fields of architecture and art, including awards, publications and exhibitions for excellence in design. Atelier FCJZ has realized architectural works in various regions in China as well as overseas, with extensive experience, professionalism and creativity.
Projects of FCJZ range from cultural facilities, public buildings, housing, office design, retail and commercial design, lab design to interior design, urban planning, strategic planning and landscape design. Besides, FCJZ has the capacity of product design and exhibition design.

Anna Matuszewska – Janik

Anna Matuszewska – Janik KREACJA PRZESTRZENI, POLAND
"My workshop is a simple translation of passion and aesthetic sensitivity into work. KREACJA PRZESTRZENI means creating interiors based on the natural and selfish pursuit after beautiful, neatly organised spaces that are friendly to its users and full of positive surprises."

Marcos Samaniego Reimíndez

Marcos Samaniego Reimindez (A Coruña, 1971), offers a refreshing vision of architecture. He is able to link natural resources and cultural heritage. His contact with European culture and design, during its formative stage in Barcelona, has accentuated his professional personality: details and the spaces have a particular goal, away from the impersonality. Before returning to A Coruña, Marcos Samaniego worked in several Architecture studios: Altis Arqs. Asociados, en Tusquets, Díaz & Asociados, and Factoria d'Arquitectura.
Between 1999 and 2000, he managed School Management Workshop Arteixo, until October of that year when he opened his own studio: MAS • ARQUITECTURA. It's here where he encourages his professional interests. His designs have been featured in international publications, newspapers and magazines. The combination of tradition and design is highlighted by experts, who consider Marcos Samaniego one of the great promises of Spanish architecture.

Darley Interior Architectural Design (DIAD)

Managing Director - Shiree Darley. As a key role player in establishing and leading the Wilson Associates South African office, she acted as Managing Director for the past 16 years. This new company is a culmination of her innovative international design experience and proven ability.
Operations Director – Caroline Dann. Professionally trained with a Bachelor of Architecture Degree; Caroline has achieved great success in overseeing international design teams with the implementation of numerous exclusive hospitality and high end residential projects.
Design Director – Amanda Elliott. An accredited Interior design professional, Amanda has played a leading design role with various design teams, together completing numerous award winning projects globally.
Creative Designer – Sarah-Jane Forman (Interior Designer). Sarah-Jane initiated her career designing the installation of projects for one of South Africa's leading exclusive residential designers; and then extended her focus to Resort, Leisure and Spa Design.

Atelier Heiss Architekten

Atelier Heiss Architekten, founded in 1997 by Christian Heiss, is based in Vienna. Today, projects are developed by a staff of about 30, under the management of the three partners Christian Heiss, Michael Thomas and Thomas Mayer.
Atelier Heiss Architekten implements individual and unique projects in the fields of architecture and interior design. In close cooperation with their clients, they develop custom-made solutions with the highest standards of architecture.

Brunete Fraccaroli

The renowned architect Brunete Fraccaroli, graduated at the Universidade Mackenzie in Sao Paulo and worldwide known as the colorful architect is a respected and requested professional, for both, house and commercial projects.
Among her most acknowledged projects are: The Glass Garden for "Espaco Deca"1999, which was mentioned in The New York Times, won the CREA Prize from Belo Horizonte, the 1st place at the Espaco D Award; The Glass Garage of "Casa Cor"2001, -project in honor to her father who loved cars-this area won the Solutia Design Awards 2002; the commercial project of the Glass Store in Sao Paulo, which won the contest Solutia Design Awards 2003 together with the Spanish architect Santiago Calatrava.

Ippolito Fleitz Group

Ippolito Fleitz Group is a multidisciplinary, internationally operating design studio based in Stuttgart. They are identity architects. We work in unison with their clients to develop architecture, products and communication that are part of a whole and yet distinctive in their own right. This is how they define identity, with meticulous analysis before they begin, with animated examination in the conceptional phase, with a clarity of argument in the act of persuasion, with a love of accuracy in the realisation, with a serious goal and a lot of fun along the way. Working together with their clients, as architects of identity, they conceive and construct buildings, interiors and landscapes; they develop products and communication measures. They do not think in disciplines. They think in solutions. Solutions that help you become a purposeful part of a whole and yet distinctive in your own right. They architect various identities identity.

Hjgher

Founded in Jan 2006, HJGHER is a design collective with specialised backgrounds in design and behavioural science. By combining their strengths, Higher is primarily about design psychology.
In 2007, they received an award from American Institute of Graphic Arts and CLIO awards for their self-published book, "Dear J". In 2008, the Asia Interactive Awards awarded them for their HJGHER website. In 2009, their interior design and branding project, Kith Café, was awarded by the Hong Kong Designers Association Asia Design Awards in the Spatial Design category. In 2010, they were honoured with top 40 under 40 designers to watch by Hong Kong Perspectives 2010.
In conjunction with the Singapore Design Festival 2009, they launched their self-published magazine Underscore, selling out all 1000 copies of their first issue in two months. A second print run of 1500 Underscore, selling out all 1000 copies of their first issue in two months. A second print run of 1500 copies were printed for distribution in Amsterdam, Auckland, and so on, which too sold out. Most recently, it was the only Singapore/Asian publication to have won a D&AD 2010 In-Book nomination for Entire Magazines in the Magazine and Newspaper Design category.

Cheremserrano Arquitectos

Cheremserrano Arquitectos is a young firm of architectural and interior design. The firm's found in 2004 by Abraham Cherem and Javier Serrano.
Architect Abraham Cherem Cherem was born in Mexico City, on November 1st, 1982.
He obtained his degree in Architecture in the Universidad Iberoamericana, in Mexico
City (2001—2006). Architect Javier Serrano Orozco was born in Mexico City, on March 22, 1982. He obtained his degree in Architectura in the Universidada Iberoamericana, in Mexico City (2001-2006).
Awards:
- Premio Nacional de Interiorismo 2008. Category: Restaurant: El Japonez
Santa Fe.
- World Architectural Festival Barcelona 2008. El Japonez Santa Fe. Shortlisted.
Cheremserrano Arquitectos offers a balanced design that respects the client demands, economics, functional use of space, and is respectful with the environment.

NMD I nomadas

NMD I nomadas is primarily a creative team of a multiscale vision whom approach every challenge with a comprehensive and strategic vision of solid concepts, covering urban project, the architectural development in their various scales and functions, landscape architecture, interior architecture and urban public transport. Their process pursued the production of unique answers to each situation, own of the coexistence of multiple environments and the search of cues that develops in thematic, political, socio-cultural, financial, economic, physical and urban aspects that ultimately require unique answers.
Constituted of over 30 professionals, they have developed more than 1,200,000 professional hours of consulting and designing; structured partnerships with important and specialist consulting firms; worked with governmental institutions, petroleum industry, corporate market, communities, allied consultants, professional collages, universities and real estate, sport-recreational, cultural, educational, sectors such as health, transport, commercial and housing.

Paul Burnham Architect Pty Ltd.

Date of Birth: 29th September 1959
Education:
Hale School, Western Australia
University of Western Australia
Qualifications:
Bachelor of Architecture/University of Western Australia/1982
Associate Member/Australian Institute of Architects/1986
Registered Architect/ Architects Board of Western Australia/1986
Registered Architect/Architects Registration Board, UK/1989
Corporate Member/Royal Institute of British Architects/1989
Employment:
Marcus Collins Architects/Perth, Western Australia/1985 - 87
Barrett Lloyd Davis Architects/London, UK/1987 - 89
Terry Farrell Architects/London, UK/1989
RHWL Architects/London, UK/1989
Private Practice:
Paul Burnham Architect/Perth, Western Australia/1990 - 2011

Fusion

Fusion, a specialist design company founded in 1998 by Roger Gascoigne and Sophie Douglas, offers a unique combination of the skills from the architectural and interior design professions. With a staff of approximately 16 talented architects and designers, Fusion aims toprovide a market-leading service by exceeding client expectation in every aspect of their input, from the inception of a project to its completion. Fusion's clients and projects are generally within the bar, restaurant, catering and hotel sectors and their success has been based on building long term relationships with, and attracting new clients with their successful and award winning projects.
Fusion's recent client/project list includes Leon, Everyman Cinemas, Fire and Stone, Rocket, Drake and Morgan, Villandry Kitchen, Nandos, Turtle Bay, TGI Friday, Balans and La Tasca.

Sergey Makhno/Butenko Vasiliy

Sergey Makhno is an artist, architect, designer, workshop project manager. Graduated from Kiev National University of Building and Architecture as well as from Academic School of Design (Moscow). His career as a designer is not an ordinary one. In his early childhood Sergey showed interest to design work, however, he started his career as a trainer of KioKushinkay karate and grew to professional writer and designer of Author's furniture. In 1999 he founded "Makhno workshop". Sergey believes that his imaginative vision is combination of contradictory backgrounds and creation of artistic mixes from seemingly incompatible things. He prefers rather not to have a rest but pay a visit to world design exhibitions and exposure into antique shops. May come up with an invention and devise not only well-forgotten old things but also new ones never thought of before. Serghii Makhno and Vasiliy Butenko designed the amazing restaurant Twister together.

Guru-Design

Web: www.guru-design.com.tw

Jean de Lessard, Designer Créatif

Jean de Lessard, Designer Créatif is a firm offering its clients complete and innovative solutions in interior planning and design. This award-winning company has seen its work published across Canada over the past 20 years, distinguishing itself for a sophisticated clientele in the corporate sector and restaurants, not to mention private residences, hotels and boutiques.
Bolstered by its savoir faire and guided by the creative flair of its principal designer, Jean de Lessard, this firm, boasting a host of specialists, is renowned for creating stylish and resolutely contemporary interior spaces in North America and Europe.

AB Concept

An accomplished design firm with an eye for the unique, AB Concept focuses on high end hospitality projects around the world. Directors Terence Ngan and Ed Ng have a gift for balancing stimulation with a sense of deep tranquility, and through it they aim to anchor each project to a client on a physical and emotional level. But while they happily push the creative envelope – no project is complete without a bold statement or a few hidden surprises – they heartily shun token decoration. At AB Concept luxury is not expressed in frills and ornamentation but rather in the perfect sculpting of form to function, the careful balancing of materials, the emotional play of light and proportion. Since its formation in 1999 the studio has developed a vision and a reputation that spans continents and cultural themes and its shelves are well stocked with international awards. By getting to the heart of a client's vision and taking it to new levels, they bring a perfectly tailored, one-of-a-kind product to the table every time.
web : www.abconcept.net

Norihisa Asanuma/Naoki Horiike

Norihisa Asanuma
1980 Born in Lwate, Japan
2004 Graduated from Department of Architecture, Kokushikan University
2004-2007 Joined Architecton (Akira Yoneda)
2007 Established "epitaph"

Naoki Horiike
1981 Born in Shizuoka, Japan
2004 Graduated from Department of Architecture, Kokushikan University
2005-2007 Joined Moriyasu Hase & Associates
2007 Established "epitaph"

BNKR Arquitectura

Bunker Arquitectura is a Mexico City-based architecture, urbanism and research office founded by Esteban Suarez in 2005 and partnered by his brother Sebastian Suarez. In their short career they have been able to experience and experiment architecture in the broadest scale possible: from small iconic chapels for private clients to a master plan for an entire city. Bunker's unconventional approach to architecture has continuously attracted attention and generated public debate with projects such as a three-kilometer habitable bridge that unites the bay of Acapulco and an inverted skyscraper 300 meters deep in the main square of the historic center of Mexico City.

3SIXO

3SIXO is an architecture firm headed by Kyna Leski and Chris Bardt. Their work is best defined by the spatial idea that is created out of the forces, content and limits unique to each project. This spatial idea acts as a guide to the many decisions through all stages - from sketch through construction. In this way, one does not perceive an applied theme or style. Instead, one experiences a sense of "place" that is integral to the situation, directly through the workings of the architecture: through phenomena-light, sound, view, shadow, balance -and through images triggered by the imagination.

Johannes Torpe Studios

Johannes Torpe Studios is a bastard love child of a company. It's a playful crossbreed of design and music infused with a bit of madness, run by the brothers Johannes and Rune RK. The studio has three core business areas. We create interiors, furniture and industrial design. We know no boundaries when it comes to scale, discipline, media or market. In our creations we include all senses and revolutionize all dimensions. We have produced furniture for Moroso, SpHaus and Hay. We have designed the high-end shopping, dining and clubbing experience that define SUBU, NASA, Supergeil, Evisu, Noir and Paris-Texas and many more. This summer we're launching Copenhagen's first cupcake bakery, "Agnes" and our new office in Beijing. Our pipeline is thick with projects that we can't wait to let you in on.

Edmond Tse

Edmond Tse is a registered architect and designer in Hong Kong, specifies in high-end residential and academic institution. After graduating from the University of Texas at Austin, he worked for international brands such as RTKL International, Leigh & Orange and Cypress International. He is now the director of Imagine Native and is also a visiting lecturer at Hong Kong Design Institute and Macao Polytechnic Institute.

Alexi Robinson

Alexi Robinson is a Hong-Kong based interior designer, who works internationally with clients on the creative direction and design of bespoke interior environments. Often collaborating with designers from related disciplines, Robinson's approach embraces all aspects of the interior to present a cus-tomized and unique brand experience. Recently Robinson was responsible for the design of Press Room Group's flagship restaurant, SML, and past clients include award-winning restauranters Can-teen, Acorn House and Soho House Group. Robinson's skills extend to design writing and exhibition design and curation, notably 100% Design London, 2007.

Daniel Scovill

Registered Architect, California, 2008
Bachelor of Architecture, California Polytechnic State University, San Luis Obispo, 1999
Collaborative design innovation has been a long-standing interest of Daniel's, extending back to his experience at Gensler, where he served on various showroom and prototype design teams for clients such as Volkswagen, Mazda, Lincoln, Ford, and General Motors. At Axis Architecture + Design, he applied his background in merging branding with built environments to restaurant and hospitality projects.
Daniel founded Arcsine in 2003 to pursue a hands-on collaborative approach to the design, detailing and construction of hospitality, commercial and custom residential projects. He holds a bachelor degree of Architecture from California Polytechnic State University in San Luis Obispo and is a licensed architect in California.

后记

本书的编写离不开各位设计师和摄影师的帮助，正是有了他们专业而负责的工作态度，才有了本书的顺利出版。

参与本书的编写人员有：Ed Ng, Terence Ngan, Paul Burnham Architect Pty Ltd., Jody D'Arcy Photographer, Anna Matuszewska – Janik, Paweł Penkala, Ippolito Fleitz Group - Identity Architects, Zooey Braun, Alexi Robinson Interiors and Michael Young, Katsunori Suzuki & Eiichi Maruyama, Diamond dining, Cheremserrano Arquitectos,SAQ, Frank Tjepkema, Janneke Hooymans, Tina Stieger, Rob Wagemans, Melanie Knüwer, Charlotte Key, Erik van Dillen, Sofie Ruytenberg, Ewout Huibers, Frankie Yu, Naoki Horiike, Norihisa Asanuma, Mitsunobu Horiike, Marcos Samaniego Reimúndez, Ana Samaniego, II BY IV Design Associates, Antonio Di Oronzo, Michael Anhammer, Christian Ambos, Harald Höller, Hertha Hurnaus, Hideki Kureha, Noriaki Takeda, Ikuma Yoshizawa, TAO THONG VILLA Co., Ltd.,Chang Yung Ho, NMD|nomadas, Farid Chacón, Francisco Mustieles, Claudia Urdaneta, Luis Ontiveros, Jim Hamilton, Renzo Mazzolini, Helen Hoghes, Gareth Gardner, Caroline Collett, Simone Pullens, Lieve Vandeweert, Koichi Takada, Sharrin Rees, Mario Gottfried, Adrián Aguilar, José Alberto Rodríguez, Javier González, Óscar Flores, David Sánchez & Héctor Hernández, Fabiola Menchelli, Zaida Montañana, Yuriy Ryntovt, Andrey Avdeenko, Sophie Douglas, Darley Interior Architectural Design (DIAD), Eyal Shoan, Oded Hagai, Iris Rubinger, Eldad David Husravi, Liza Grishakova, nemaworkshop, Philippe Starck, David Rockwell, Blandon Belushin, Kyna Leski, Christopher Bardt,, John Horner, Namgoong Sun, Urbantainer, Jorge Luis Hernández Silva, Diego Vergara, Lorena Ochoa, Carlos Díaz Corona , Studio Kuadra, Photophilla, Sergey Makhno, Butenko Vasiliy, Joshua Aidlin, David Darling, Roslyn Cole, Adrienne Swiatocha, Shane Curnyn, Matthew Millman, Jean de Lessard, Stéphane Proulx, Jean Malek, Jeroen Vester, Ulrike Lehner, Femke Zumbrink, Hjgher, the Shooting Gallery, dan pearlman Erlebnisarchitektur GmbH, Yuhkichi Kawai, Toshihide Kajiwara, K-Studio, Yiorgos Kordakis, Pascale Gomes-Mcnabb, mark roper, sharyn cairns, Tim Mutton, Jo Sampson, Jordan Littler, Jaime Gaztelu, Mauricio Galeano, Iñigo Esparza, María Asún Osés, Pablo Ausucua, Arcsine Architecture, Sharon Risedorph, Edmond Tse, Atelier Heiss Architeckten, LW Design Group, 3six0 Architecture,Brunete Fraccaroli, Yasumichi Morita, Neil Hogan, Brendan Heath, Peter Burgstaller, Klm-architects, Johannes Torpe, Bruce Buck, Younjin Jeong, Daesoon Hwang等。